THE WAR

361 DAYS, 12 HOURS
AND 27 MINUTES IN VIETNAM

THE WAR

GERALD A. SPENCE

KS
Kravitz & Sons
INNOVATORS IN PUBLISHING, MARKETING AND ADVERTISING

Kravitz and Sons LLC
1301 Farmville Blvd, Suite 104
Greenville, NC 27834

Published by Kravitz and Sons LLC.

ISBN: 979-8-89639-307-8 (sc)
ISBN: 979-8-89639-306-1 (e)

Library of Congress Control Number:

Contents

The Draft

It's 1966 and I, Jerry Simpson, am a twenty two year old married male. I have just completed my twelve week stint in rookie firefighter school and assigned to a Fire Company, Engine 19. I have finally realized my life's ambition. Since early childhood, I had wanted to be a firefighter. Most little boys grow out of that dream. I never have.

I work a somewhat confusing schedule, quite different from a normal or routine work week. I work three days of ten hours and three nights of fourteen hours and get three straight days off. The work, if you want to call it that is exciting and everyday brings a new experience that no ordinary job could ever come near.

My beautiful wife Sherrie, whom I married in 1964, works for the Federal Government. She has loved me since we met in the tenth grade of high school, though I will never understand why. Life was wonderful for us, not true for our country.

The United States is engaged in the Vietnam thing and President Lyndon Johnson has decided to escalate our involvement in order to eliminate the aggressor forces that are attempting to overrun South Vietnam. In order to provide sufficient manpower for this effort, he has implemented a new draft initiative "calling up married men without children". This has caused some of my married friends to become very tentative about their future as a civilian. Not I of course, because of my draft deferred profession, there is no danger of being called up. At least that is what they tell me.

One night after my shift I returned home to the apartment Sherrie and I had recently moved into. I went through my routine of checking the mail, opening and closing the refrigerator door and lighting up a Marlboro. I skimmed the mail and saw a letter from the Selective Service. Must be confirming my deferred status, I thought. The heading read, "Greetings from the President of the United States." Enough of my friends had already told me what followed. I was going to war.

I called Fire Department Headquarters and spoke with the personnel office. They advised me to write to the Draft Board and ask for a two week deferment, while they worked on my status. I did as directed and got an extension and reassignment to the city draft board. The events that followed were a blur of requests, rejections and dead ends. But ultimately, I had to go in the military as a draftee or volunteer for service. I was assured I would have my old job waiting when I returned. Hey how bad can it be anyway, lots of guys have done it?

On, September 28, 1966 I reported for induction into the Army of the United States. We loaded onto buses and headed to Fort Bragg, North Carolina. I celebrated my twenty third birthday a few weeks afterwards by scrubbing pots and pans on KP. What a miserable situation, home sick, unsure of where my life was going and wanting desperately for this to be over with. I was the oldest draftee in my entire Basic Training Company. The First Sergeant, Company Commander and Platoon Sergeant are the only ones older than me.

Basic Training ended around Thanksgiving 1966. I reported to Ft. Gordon, Georgia for Advanced Individual Training. Shortly after that I went home for a week of Christmas leave.

Then it was back to Ft. Gordon. About fourteen weeks later, we finished our Military Occupation Specialty training. I was reassigned to the 337th Signal Company at Ft. Bragg, NC. The Unit was being staffed for deployment to Vietnam. We completed jungle warfare training and all that stuff and were scheduled to ship out in July or August of 1967.

The one bright spot in this whole series of events was that Sherrie and I found an off base trailer. I got thirty days leave before our departure.

During those thirty days we tried to ignore the immediate future and concentrated on the present. We visited friends and some family members. It was hard to visit with my father, who was a shell of the man I had left only several months ago. My siblings were already decimated by the dissolution of our family. My mother remarried and had the two youngest with her. I did not see them or her at all.

In July Sherrie told me she was pregnant. My emotions swung from elation to apprehension. What if I didn't come back or worse yet was crippled or screwed up in the head? How could a young woman contend with those circumstances, especially with a small baby? It was most difficult to say goodbye to her. I wasn't sure if it was a short term or forever departure. "I love you," did not seem to encompass my feelings, nor, "I'll miss you," or any of the platitudes we normally speak when going away. And men, real men don't cry, at least not publicly and not yet.

Making the Journey

It's August 1967, I 'm at Fort Bragg waiting for the word to move out. The skinny is 0500 to Pope Air Force Base, contiguous to Ft. Bragg and the first part of our adventure and new world experience. The buses arrived and we boarded with our weapons and gear in hand. Nervous butterflies circled in our stomachs while unanswered questions flitted through our minds. On the runway were the charter jets that would carry us to San Diego and the waiting merchant marine vessel the Nelson M. Walker. We took off, refueled at Will Rogers Airport in Oklahoma and arrived in sunny California late that afternoon. Another bus ride and we reached the terminal. The ship was glib, gray and gloomy. The bilge pumps were at full force, pumping out the Pacific as fast as the water seeped into the hold.

After the normal military waiting period (seemingly forever) the companies assembled and were escorted to their assigned berthing areas. We dropped our stuff, stowed our gear and settled into our canvas fold down cots. About 4200 of us planned to take this eighteen day trip. Once out on the Pacific, the ship rocked and rolled. Each wave seemed to hold the vessel suspended for a brief, but scary, few seconds, then tumbled into a valley of water and spray.

Guard duty, KP and other assigned duties awaited us at sea. I got KP right at the start, six nights in a row, midnight to 0600. I carried crates of potatoes and cartons of eggs from the bow to the galley. It seems like three steps forward and two reeling backwards. I frequently stand under a vent stack that feeds huge volumes of fresh air from the surface. It keeps me from losing the little edible food I consume. Thank goodness the Merchant Marine maintained a large stock of ketchup, also known as "navy gravy." It kills the taste of everything effectively.

We eat in shifts. Each group has thirty minutes to get in and out of the mess area. Everything except the food is made of gray metal or stainless steel. Thick white bowls, cups and plates withstood the clumsiness of the GI, the tossing about of the ship and held our alleged nourishing meals. After the meals and with no work details there's nowhere else to go but topside, we sit around on deck. No one was allowed below deck until 2100 hours and then lights out was at 2200.

Bathing and hygiene opportunities are also a real competition. The first one hundred or so get warm, fresh water showers. After that, pure ocean salt water was the only source available. I decided, after my first attempt, never to use soap when I take a salt water shower. When combined it makes a gooey substance, resembling a jelly fish. It just sticks to your body and is very hard to rinse off. Some folks waited several days to experience even that. Yes, it was a bit raunchy.

It's the thirteenth day of our voyage and we stop in Okinawa for provisions and are given an eight hour pass. There, of course, were conditions we had to adhere to. We couldn't go into the city. We had to stay on the military compound and we had to return to the Walker by midnight.

When the ship arrived at the harbor, we all stood out on the deck. The tugboats brought us near a military band that was posted at the pier. They played the routine patriotic songs and marches it was their job in support of the war. I estimate

4,198 of the 4,200 of us were more focused on getting onto land, any land, than listening to them play. I made a quick trip to the head, changed into khaki dress uniform and went on my way.

Buses lined the dock area. We charged the vehicles like a swarm of flies descending on donuts. We overtook them in a light brown wave of humanity. The trip to the installation was brief. The Enlisted Club was huge. I guess it was designed for one specific purpose, to entertain GIs who were going to Nam or those who were on the way home. Loud bands played in the several ancillary rooms. We grabbed a beer, ordered a steak and drowned our anxiety in the moment. The bands were mostly semi professional groups from the States hired by the USO. They knew all the latest tunes and, though not the real rock stars, did a great job of taking the edge off as we got wasted. By 2000 hours, we were quite drunk and emboldened by the suggestion, "Let's dump this place and go downtown."

We edged out of the club, hailed a scooter cab and instructed the driver to get us off the base and into town. It is amazing how much five dollars can buy. It wasn't very difficult to get away. In fact, we could have just paid him two dollars for the ride. We made it into town and strode into, actually stumbled into, the first neon lighted "American Bar" we saw.

Immediately, we recognized some Walker GIs. They bought the first round and we bought the second. Soon we lost count. Later, three Marines who were permanent party in Okinawa walked in. We start talking and of course the conversation turned. We asked, "Have you been there? What was it like? How many of their friends were wounded or killed?" As we listened, reality surfaced above my intoxication. I was at death's door, just waiting for the grim reaper to snuff me out.

They related similar stories to those we had heard from some of our Drill Instructors and others we had talked with stateside. You didn't know who your enemy was, where he/she would spring their attack on you, what weapons they might use and you learned not to depend on any of the nationals. Your friends during the day could turn into your killers at night. The Vietnamese people (RVN) were ambivalent about fighting, and not too charged up about winning anything, even freedom from the North.

A beer bath and brawl started at about 2300 hours. This dramatically interrupted our information gathering. We relieved our suppressed anxieties by engaging our fellow soldiers in corporal punishment. People yelled, sirens wailed, guys bled and we all ended up in the custody of the Military Police (MP). After shelling out fifty bucks (each of us paid the fine for damages to the bar), we were carted into the waiting deuce and a half trucks which carried us back to the ship. Bloodied, khakis ripped and dirty, many of us displayed varied layers of vomit. We were a sight.

The MPs offloaded us. We were supposed to line up in formation together so our Duty Officer (DO) could claim us and usher us on board. We staggered and dragged onto the ship and toward our sleeping berths. I was still throwing up my consumptive delights into whatever area my face was pointing. So, someone assisted me to the head and placed me on an available one-holer (latrine). The one-holers lined the bow of the ship, about fifteen across in three rows elevated by a slatted wooden walkway with two steps to the metal deck. To get to them I went past the showers and the urinals, which occupied the bulkhead wall on both sides of the bow.

I think it was about 0300 hours when I awoke. My throat was raw, my stomach churned and I was numb from the waist down. I must have passed out at the latrine which cut off the circulation to my lower parts. The feeling slowly returned to my legs so I proceeded through the head and located my bunk. Thankfully it was the middle bunk the guy assigned below was on KP and the guy above didn't drink. I didn't have to sleep with any foreign substances in my bed.

I was rudely awakened from my very brief nap at 0600. Everyone was ordered topside except those on clean up detail. I went to the mess hall, grabbed two slices of toast and a cup of coffee. I walked up to the deck and looked for a familiar group. Dawn was fresh, the sun slid up the skyline and life seemed wonderful. It was strange, but peaceful. Guys joked, played cards, shot dice, read and write letters. Not a bad way to pass the time.

Just how many were in that bar fight, I ask myself as I looked around. From where I sat I saw about a hundred or so guys are bruised or cut or both. I didn't see anyone with broken bones. That's good. If any serious injuries occurred we could be court martialed. Fifteen days ago most of us had not even attempted to befriend one another. Now I searched for one or two guys I could trust with watching my back and getting me home. Richard and Don fit the bill as they were also with me at Fort Bragg. I move over to the railing and start talking with them.

The morning warmed and afternoon gave way to night. The temperatures are into the high eighty's and no rain. We get ever closer to that mysterious place we heard so much about. Inevitably our conversations keep going back to the collective concerns: what will happen to us, will we all come home alive? We have three more long days of sea travel before we would see land again. This land however, I was not sure I wanted to see.

Landing in Da Nang

Three days later just after breakfast (a loose term for sure on this boat), we are on the deck and I hear someone say, "Look there's land ahead. Is that the Nam?" We all stood and strained to see what direction he was looking in. Yep, there was terra firma, getting bigger as we headed towards it. I felt the blood pumping, excitement, fear, apprehension; or maybe a combination of all three. The PA came to life and instructed us to go below deck to our company berthing areas for instructions. I knew we were beyond the tropic zone, but it was hot and sticky, especially huddled up with nearly two hundred guys.

We were off the coast of Da Nang and would be nearing the port entry within thirty minutes said the XO. He was an ROTC Second Lieutenant, fresh out of college, barely twenty one years old. Our First Sergeant was the only veteran of Vietnam. He expected to get his next stripe while here and retire when he got home. He had twenty eight years and three wars under his belt. Well two wars, Vietnam wasn't a real war, so they told us. I figured if they were trying to kill me and I was trying to kill them, it was a war.

Captain Sandifer arrived and told us the process: we'd enter the harbor by tugs and drop anchor. Soon after we would assemble by Company on deck, and offload onto LSTs, and be transported to the beach. The landing boats would ferry us to the beach and then return to the ship until all the troops are off loaded. Major news organizations will record our landing. They tie it into the election message for President Tieu's re-election effort. One other thing, we won't have any live ammo, too dangerous! We were to disembark the LST vigorously and loudly.

The ramps dropped on the shallows and forty two hundred of America's finest unarmed fighting men stormed Da Nang Beach in staggered assaults (as the LST's arrived). Fortunately we did not suffer one fatality during this landing. So the message was: war zone, no bullets, be loud and aggressive, and convince the natives to cast their vote for Tieu.

As soon as we were out of the cameras range we loaded onto deuce and a half trucks and were transported to the 37th Signal Brigade, located beside Da Nang Air Base. We could actually touch the buildings from either side of the deuce and a half. I thought to myself, one grenade per truck and 20 of us are dead.

It was a slow and somewhat winding trip to the 37th compound. As we turn the last corner we see the airfield on the right. Large aircraft hangers towered above the security fence. I lean over and shout to the driver, "What are the long green boxes that are stacked in between and taller than the hangers themselves?" He replied, "Those are caskets man, thousands of them." It was then that the reality and seriousness of my present circumstance became personal. We turned into the 37th Signal Brigade compound and made a hard right that took us to our temporary home. The convoy came to a halt and so did the breeze. It seemed that the temperature shot up to practically unbearable! Our fatigues suddenly looked like we were in a heavy rain; our bodies were actually the rainmakers.

Once off the trucks, we formed up as Companies and then divided into Platoon groups. Our Platoon Sergeant advises that we would be building bunkers while others set up tents, dig trenches and set up the perimeter. Though the entrance to the compound was fenced, I noted there were no barriers at the rear area. I saw farmers planting and tending their rice fields. They pretty much ignored us altogether.

Building bunkers was a real adventure and quite a learning experience. We were fortunate that a mountain of sand was piled up dead center of our Company area. So, about half of us started filling sand bags while the remaining guys started digging a large hole. For the first several days we filled sandbags, dug holes and placed sandbags in piles around holes. We left weapon ports in the sandbagged walls, covered the tops with plate steel sheets and then covered the steel with more sandbags. I could not estimate the numbers of sand bags we filled. I am surprised there is any beach area left in South Vietnam by the time we complete our construction.

For the first few days we live out of our duffle bags. Then the storage containers are delivered. By that time we have our own tents with bunks and mosquito netting. We retrieved our footlockers and another duffle bag of personal items from the containers.

We have been going pretty hard up to this point and the CO decides to give us a break and let us unpack our stowed gear. In eight days we have transformed that flat area with a mountain of sand into a neat row of olive drab green tents, trenches and a defense perimeter of concertina wire, claymores and bouncing Betty's. Our rear area had five neatly spaced bunkers that would provide safety during mortar attacks and protective shelter with firing positions for us.

We had guard duty for two hours on and four hours off every other night. This was in addition to working the day of and day after. We didn't have much time to do anything other than write letters, and try to nap when we could.

An enlisted club was at the other end of the compound. Most permanent party locations had one. There we could get a cold drink, play cards or darts and listen to the juke box (for those unaware, a juke box was a large self contained record player that held 40 or so records), never mind you will want to know what a record is next.. Some of us had battery powered portable radios and we listened to Armed Forces Radio. It was a good thing and helped me feel less detached from family and friends back home. And I missed home. Just before I shipped out, I found out my wife was pregnant. This will be our first baby and I won't get to experience this pregnancy with her, I regret that.

The XO walked into our tent. We learned through our training that snipers prefer officers and senior NCOs as targets. So we didn't have to jump to attention when an officer approaches, this protocol is relaxed in a war zone. The XO is in his early twenties, an MIT engineering graduate. His name is Fred but we call him Lieutenant Derf, (not to his face). Derf is actually Fred spelled backward. He said the NVA radio station had Hanoi Hanna on their broadcast earlier this afternoon. Apparently their resources were pretty good. They already knew which Companies were on our ship, the ship's name and where each unit went after disembarking.

He also said that she stated that some of us would be getting a welcoming tonight, specifically the 337th Signal Company at the 37th's Compound. He wanted us to be sharp and keep our eyes open. We had already experienced some incoming rocket and mortar fire due to our proximity to the air base. The VC tried to knock out aircraft as they took off and landed. We were close to the end of the primary runway and they usually missed the planes and the projectile landed near us. When Lieutenant Derf leaves, we resumed our casual conversation.

One of the guy's says, "Frost is getting sent home and likely a medical discharge." "Why, what's wrong with him?" we ask. "Apparently the anti malaria drugs and the oppressive heat left him unable to sweat. His body can't cool him off and he started swelling up due to build up of fluid. They are also giving him a Purple Heart for 'war related illnesses' or some such reason."

It was my night to refuel the 25KW generators that provided power for the tent city. The fuel tanker pulled up to the fuel trailer and I opened the top cap, inserted the fuel hose and told the driver to proceed with the transfer. Fueling complete, I checked the gauges, entered the readings on the log book and returned to the tent.

The night seemed a lot darker and foreboding as I headed back. It had been a long day and I was ready for some sleep. When I returned, the four New Yorkers were playing cards and arguing which one is the most New York-ish. My money was on Cherico for sure because of his accent.

Second place went to Jesus Medgarez; Puerto Rican English is quite unique. The other two Anton Brown and Edgar Wilson were both black guys from Harlem and more conversationally eloquent than the other two clowns. We were definitely a homogenous group of guys, every color and ethnicity you can imagine

I found out some reenlisted after leaving another branch of the service. It's hard to imagine anyone coming back for more of this. I figured it took all kinds to fight a war. Some of these guys had been together since basic training. A couple of guys from my Advanced Individual Training (AIT), at Ft

Gordon were here, neither was in my platoon. As the night progressed, things quieted down. I crawled into my bunk, lay on my sleeping bag, and tied and zipped my mosquito netting. I closed my eyes for another night.

Rockets and Buffaloes

2 6 August 1967, Was I dreaming? No. Guys were screaming, "Incoming!"

I heard a sloshing sound, a thud and then the explosion. What was going on? My mind cleared and I pulled the zipper back, rolled to the floor and grabbed my flack jacket and helmet from under my bunk. I snatched up my weapon and ammo belt. With simultaneous movements, I low crawled as quickly as my body allowed and reached the twenty two feet from my bunk to the bunker at running speed. I think I matched my best running time even though I crawled. I just kinda rolled, lunged and poured myself into the opening. Someone just as anxious might have rolled right in after me so I moved fast to avoid a pile up at the doorway.

Three of us made it in here so far. We positioned ourselves at firing points along the bunker and looked for intruders along the perimeter. I struggled to get my mind in order. My adrenalin pumped as the incoming rounds (the same type they fire at the planes over at the airfield) descended near our location. They came in rapidly for only a few minutes. The VC tactic was to strike fast and run.

Sergeant Dower, our Platoon leader slipped into the bunker for a count, "We won't be going back to bed tonight. What time is it anyway?"

It was 1120 hours. I was only asleep for an hour! Dower crawled out and headed for the next bunker. Cherico slid in with us. Wilson and Brown followed soon after. They took their gun ports and we peered out at the darkness. The guys in the mortar pit were getting ready for illumination flares. Then I heard a snap as the flare ignited. An eerie reddish white light spread over the horizon. It reminded me of the last night of combat training when we did a live fire exercise through an obstacle course.

Cherico spoke first, "Damn! Those bastards nearly killed my ass!"

We all felt the same rage, but instead we laughed and tried to act tough. While I chuckled, I shook uncontrollably in the darkened bunker. The field phone rang and startled us all. It was the commo bunker asking for a check and headcount.

After a while the firing ended, no more incoming. It didn't appear we were being attacked by ground forces. As things slowed down, I noticed I was getting cold even though I was drenched in sweat and the temperature in this bunker was very warm. I concluded the chill was from fear and excitement.

As I pondered this, I heard Jesus cussing and yelling for us to give him a hand. I looked out the bunker opening. He screamed, "I tripped over the damn tent ropes and fell into the drainage trench between the tents. I think I broke my damn leg."

We pulled him in as he yelled in Spanish. Wilson cranked up the phone and reported Jesus' injury. They instructed us to keep him comfortable. They will retrieve him at first light. Later Sergeant Dower returns and confirmed that Jesus' leg is broken. Then to our amazement he say's, "Well soldier, you will be the first combat related Purple Heart recipient from the 337th Signal Company in Vietnam. Congratulations!"

"Wait. Whoa. Slow down here. He's not shot," exclaims Cherico, "How's he rate the Purple Heart?" Simple, the injury occurred during an enemy attack which leads to a temporary debilitating ambulatory period, it's all you need. He will be going down to Cam Ran Bay for treatment in air conditioned comfort for the next several weeks." When Dower left we teased our injured comrade until dawn cracked the darkness.

Precisely at first light the medics retrieved Jesus. We will never see him again. I understand that he was assigned to a unit in Nha Trang after his leg healed up. Little did we realize that he would also be the "only" medal recipient from the 337th.

We saw the perimeter quite well now. The natives started their day as normal, working the fields without even a hint of last night's events slowing them down. I wondered if any of them were involved in the attack. We got the all clear and emerged from the bunkers tired, hungry and really needing to pee!

Back at our tent, we surveyed for damage and found nothing except a few small holes from impacted rockets. We were on heightened alert for the next few days and rotated the bunkers every two hours, starting that morning. We were assigned an M-60 and two M-79 grenade launchers for each bunker. Since I was the oldest guy in the tent I was assigned as the M-60 gunner spot and that sucker was heavy. To make matters more interesting, Cherico was my assistant. He carried the tripod and ammo boxes. Wilson and a Chinese kid named Chin Lee, (his parents were potato farmers in Utah), were assigned the M-79's. It occurred to me that we survived our first warfare experience. I felt confident and cocky. I had no idea what was coming next.

It had been a month since I boarded the Walker and four days since the "Welcome Ceremony." We were still doing the two hour flip over stint in the bunkers. It seemed each day we spent in the "hole," the more repugnant the smell. It was probably ninety percent humidity and at least that same temperature or higher. I marveled that our jungle fatigues could dry out just in time for us to go back and repeat the sauna experience. The nights weren't that bad and other than really wanting to fall asleep, bunker duty was cooler and went by faster, it seems that way anyhow. We learned to stay alert and close at least one eye when we lit up a cigarette. The eyes acclimated to the darkness and any bright light caused us to lose acuity for a brief period. The mortar team was still firing illumination flares. We also closed out eyes when we heard them pop. I wondered why they wouldn't issue night vision viewers like the ones we used stateside during jungle warfare training. I guess they were too expensive or in short supply.

I left the M-60 in the bunker, set up for action. All we needed to do was load ammo into the receiver. We all had M-16s issued to us. It was a fine weapon, just not always functional. Most problems were due to operator negligence or jams from rapid fire. We got those weapons while at Ft. Bragg prior to shipping overseas. We had M-14's before then and during basic training. They were heavier and bulkier, accurate and potent with the 7.62 mm projectile. The M-16 could do a lot of damage to the victim upon exit. I shot Sharpshooter with the M-14, scored Expert with the M-16. If we had to re-qualify on the M-14, I probably could have done Expert on it too.

The local farmers and their families were out beyond our perimeter, plowing the paddies, planting the rice and irrigating the plants. It seemed so peaceful, except for the planes and choppers taking off. It was a normal day until dusk approached. I was just back from the mess tent and feeling good, time to relieve the guys in whiskey three. The two guys were ready to get out, hungry and smelly. One guy, Rooster, was a trucker from Missouri. He hauled chickens and feed mostly. His side kick was from Illinois. We called him "Gomer." I'm not sure if the show about Gomer Pyle was on then. If not, they certainly wrote it around this guy. He for sure was at least two cards short of a full deck and always asking the dumbest things.

We exchanged rude remarks about one another as we took our respective places. None of us watched outside. Whiskey two opened up with M-16 fire. I slid to my M-60 port and saw a large dark object headed for the concertina perimeter. We locked and loaded. Rooster said, "What is that?" "Beats me," I said, "but we can't let it into the perimeter." We loaded the ammo belt into the M-60 and opened fire. In seconds the field phone came to life. Gomer answered it yelling, "It's VC (Viet Cong). They are all over us!" Then all five bunkers were in rapid fire on the intruding force. Suddenly, it disappeared and we stopped firing. An illumination flare went off and we saw it just ten feet from our barrier. Our intruder was a large water buffalo now in its final throes of death.

Everything went quiet until we collectively say, "Oh Shit!" The field phone rang again. Nobody wanted to pick it up and report we were attacked by a water buffalo, that we expended hundreds of rounds of ammo on a water buffalo, an unarmed water buffalo.

Rooster says to Gomer, "Okay dumb-ass tell them about the VC (Viet Cong) you shot."

Platoon Sergeant Dower couldn't stop laughing. He attempted to control his voice when he answered the field phone. He explained, "Apparently a crazed domesticated water buffalo charged the location. The troops opened fire to prevent it from entering our perimeter. They succeeded in their efforts and the attacker has been subdued." He told Gomer that he'd probably get a Bronze Star for his bravery under fire. Throughout the night we heard sporadic laughter from various areas within our tent city.

The next day our XO (Executive Officer) sent a detail to retrieve the water buffalo. He invited the occupants of whiskey three to do the honors. We trucked to the site with a wrecker and an open bed deuce and a half. When we arrived, an agitated farmer and his crying children greeted us. The farmer, via an interpreter, demanded payment for his 600 pound loss. We would discuss this later.

We looked at the huge carcass. Rigor mortis had set in. The wrecker driver handed us two long canvass straps. We laid them out rolled the animal onto them, wrapped them around and hooked them to his crane. It took us about three hours to get that sucker secured and on the crane. Then we escorted it back to the 37th Battalion compound. Upon entry our fellow troops were most ebullient in offering comments and catcalls regarding our significant accomplishment.

The CO (Commanding Officer) and First Sergeant arrived; counted the number of hits they could identify on this creature and announced at least four hundred and twenty three hits out of 700 verified rounds fired.

Our CO (Commanding Officer) later announced that the U.S. Army was unwilling to pay the farmer. Therefore, we each had to pay. A good used water buffalo went for about five hundred American dollars. We could either pony up the money or he would deduct it from our pay. That water buffalo cost me twenty bucks.

Detail to Nha Trang

About two weeks after the water buffalo incident, 12 September 1967, the CO (Commanding Officer) gave us a positive accounting of what a talented and unique group of soldiers we were. Then he told us that the Company was being dissolved in order to replenish units with manpower deficiencies to bring them up to strength. Some of us would be going to Chu Lai for the formation of the American Division. Others would be going to various other locations as required. We wondered what kind of units, signal companies, engineer or infantry or armor units. "I cannot tell you what you will be doing," he said, "but I am sure you will do your best." The groups were already selected and some would depart the compound as soon as the next day.

Rooster, Wilson, Lee and I were the second group out. We were reassigned to the 518th Signal Company in Nha Trang, Central Highlands here we come. At least we would be out of the jungle in a relatively quiet area. At 1030 we said our brief goodbyes and loaded onto a jeep headed for Da Nang airbase. There we boarded a C-130 airplane. The plane moved on to the runway, revved up the engines and was on the way. Our flight was pleasant and relatively short. After a smooth landing a buck sergeant with a three quarter ton truck hauled us to our new Company.

This area is quite a contrast from the seacoast we had been at. There are no expansive rice paddies, but lots of farms, cattle and all sorts of worship places, Catholic, Buddhist and others I do not recognize.

The South Korean Army is significant in this city and is seemingly at home here. The 518th site is located inside a fenced compound with about fifteen masonry buildings with metal roofs, mesh metal windows and concrete floors.

The Company office was our first stop. The First Sergeant welcomed us and said we didn't have long to wait for our next departure. We would replace a group in the Central Highlands, on an outpost named Pre Line (Pray Lean) mountain. An MP (Military Police) Company and a contingent of Montagnards were already there. (I later learned that the Montagnards were indigenous, largely Christianized tribal people from Vietnam's central highlands. They were friends of American soldiers during the Vietnam War. According to very reliable resources, more than 50 percent of adult Montagnard males were killed alongside American soldiers during the Vietnam War.)

At this point I thought, "Mounta what? What the heck is that? And what do MP's have to do with Signal Corps soldiers?"

Until then our only exposure to MPs was our stopover in Okinawa. We did not necessarily welcome working with them. The First Sergeant answered our unasked inquiries, "Now fellas, all your questions will be answered in the morning during the briefing by Lt. Mac Cauley, our new XO. Sgt. Buckman will take you to the transient barracks and you can settle in, chow call is at 1730 hrs. The PX (Post Exchange) will open from 1500 to 1930 in case you need to stock up on stuff."

Wilson and I headed out to explore the place. The PX turned out to be about the size of a small bathroom and only two people could enter at a time. We tired of waiting in line and headed back to our billet. As we approached the building Wilson said, "That's a familiar sounding voice, sounds like Sgt. Dower."

"Naw," I say, "we left him back in Da Nang."

Sgt. Dower greeted us at the doorway. He will be our NCO (Non Commissioned Officer) and is going with us to Da Lat. "Whoa Sarge," I say, "we ain't going to any Da Lat we're headed for Pre Line Mountain."

"Yeah, yeah," he says, "I know. But Da Lat is why we are going to the mountain. Da Lat is where the Vietnamese Military Academy is located. The only road that leads in and out goes through the mountain. Our jobs will be security of the roadway."

"Wait; hold on, how do UHF (Ultra High Frequency) Microwave radio repairmen relate to highway security?" I knew the answer but I wanted someone in authority to tell me I was going to die guarding asphalt in the middle of the Central Highlands of Vietnam. "Sgt. Dower, we ain't fighting men. We haven't done any advanced infantry training."

"You killed a water buffalo didn't you? If you can shoot it, you can shoot Charlie. Don't worry; there will be plenty of people to help you out. Hey, I'm hungry, let's go over to the mess hall and chow down."

We would learn all the Army wanted us to know in the morning. We ate and walked back to our temporary, very temporary billet and crashed for the night. At least we didn't have guard duty. It had been a while since I slept the entire night through.

We woke to reveille piping through the speaker. I rolled out of my sleeping bag headed for a shower, turned on the faucet and felt warm water. Wow! I could get used to this. On our way to the mess hall Sgt. Buckman, told us to be in the Company office in an hour for a briefing. That left plenty of time to eat and get there. Breakfast was eggs, bacon, and canned pork sausage links, tough and chewy pancakes, toast and the infamous SOS (gravy with ground beef). We washed it down with coffee, strong and hot.

We filed into the Company Office and the First Sergeant introduced the XO, Second Lieutenant Mac Cauley. He was red all over, red haired, red faced. He was a Texan, an Irish Texan. He briefed us on the mission, which was unique due to the importance of the roadway and access to the city of Da Lat for obvious reasons. We would be flown into an airstrip on the top of a mountain just outside of Da Lat. There we would convoy to Pre Line Mountain. There was a total staffing of ninety two GI's and about 125 Montagnards. Most of these men were seasoned fighters and since their families were there too we could expect them to be quite fierce when under fire. The location was surrounded on two sides by tea plantations and open forested areas on the other two. The terrain was steep and heavily protected with triple layered perimeter defenses.

Our flight would leave at 0900 hours the next morning. I used the time to write home to my wife and my parents. My folks separated two days after I got drafted. Twenty six years and eight children, then it was all over. My Mom was sneaking out on Pop. He ran into her and her date one night. He didn't say or do anything then. He waited for her to come home, knocked her around and she left him with four kids under sixteen still at home. I heard from my married sister, while in Basic Training, that Dad was taking his sorrow out of a vodka bottle. This was evident when I visited him just before shipping out. I loved them both, just couldn't let the situation bog me down. The divorce was final, not much I could do. I hadn't heard from home in a month. I hoped that baby was holding on. I wondered if it was a boy or girl. Just thinking of being a father boggles my mind. What kind of a parent will I be? I know how I was raised and I wonder if my parenting skills will reflect those of my folks. There are no assurances that my desire to be a loving father will make me one. My own parents believe in corporal punishment. Having been on the receiving end of that philosophy, I hope it is not adopted by me. Don't think that I was abused or unloved, on the contrary, both of my parents worked in order to provide my siblings and me with more than they ever got. I guess the biggest thing we missed out on was the opportunity to be a family. Never really having the closeness and intimacy of sharing our joys and disappointments with our parents was something we missed. Dad worked at least two jobs and my Mom had to care for the household as well as hold a full time job. Oh well, this is not the time for problem resolution or home sickness I thought and then I dozed off.

The Caribou Flight

Reveille! How bizarre, in a supposed war zone we are still awakened to reveille and fall asleep with Taps. Its morning and the start of a new adventure, we grab some breakfast, gather our gear and report to the CQ who directs us to load into a couple of jeeps. The airstrip was just down from the compound. We came in on a C-130 but I didn't see any of them there that day. Instead I spotted a couple of Huey's (UH1 Helicopters), some Spotter/ FO Cessna's (aircraft) and what looked like an old WWII Mitchell Bomber. We pulled up to the "Bomber". It had two piston engines and was primarily grey-green with a hint of aluminum skin showing through, the Army named it a Caribou. We had a better name for it.

The back end opened like the C-130's did. An Army guy with Warrant bars strolled down the ramp. He said, "Load up guys. We will be taking off in ten and be sure to cinch up you web belts. Put your gear in the storage boxes in the center aisle, nothing lays loose."

They did the pre-flight checks and attempted to engage the lift gate, the wrenching and screaming noises came to an abrupt halt about two thirds shut. The co-pilot walked up to the cockpit, about three steps up from where we are sitting, and told the pilot to turn the engines over. The pistons slammed and black smoke shot out of the exhaust ports.

The whole plane shook like it was going to tear apart. The pilot yelled, "Sit back and enjoy the flight guys!"

We started to taxi to the end of the runway. The engines revved as the pilot released the brake. Our bodies jolted toward the still ajar cargo door. I think the pilot hit every hole and depression until liftoff. Finally we were airborne. I hoped and prayed we'd arrive at Da Lat with most of the parts of the plane still attached.

While we were in the air, our co-pilot suggested we take our helmets and place them under our butts. Small arms fire could penetrate the craft and he didn't want anyone losing the "family jewels." He also reported cloud cover over the landing strip at Da Lat and warned we might have to circle around for a while until it cleared.

The noise from the plane was deafening. We were forced to yell and often needed to repeat ourselves to carry on a conversation. Finally we resigned ourselves to the ride and looked out the porthole windows. Wilson yelled something and pointed to the window. We all turned around and saw the small dirt trail on top of a flattened mountain. It went from one edge to the other and trailed off to a sheer drop on either end. At that moment I got a nervous twinge in my stomach. This couldn't be the landing site, could it? We made a strong turn and headed into the wind. The pilot throttled back and dropped the wheels.

Though we couldn't see the approach directly, we felt the wind buffeting the plane as it eased to the ground. "Please stop," I whispered. We did. Safe on the ground, the co-pilot dropped the cargo gate and ushered us off the craft. Chin Lee turned toward the front of the plane and exclaimed something in Chinese. We saw him point out the nose wheel which was about two feet from the edge of the drop off.

The Warrant said, "Any landing is a good landing if you can walk away. Have a great time here fellas and be careful." He turned, walked back up the ramp and closed it, well almost. The plane turned, revved and started down the strip, just as the nose went over the edge it pulled up and was airborne.

"That's amazing," Rooster said, "Them Army pilots are hanging some large "kinnards". I wouldn't want their jobs, not a chance."

The plane faded to a small speck. We directed our efforts to locating transportation to the Da Lat compound. The only vehicles we saw were a fuel truck and a jeep. Sgt. Dower walked over to the fuel truck and spoke with the driver. He returned to inform us that the driver was from the Da Lat compound and will check on our ride when he goes back down. It was about a half mile to town and about two miles to the installation itself.

It was noon and we were getting hungry. It was decided we would walk down to the small village at the foot of the mountain to get something to eat. We haven't done any living on the economy yet, but this might be the start. Sgt. Dower authorized us to stow our gear in the tool bunker and head for the town. We carried our M-16's, which did not concern anyone. Most of the RVN (Republic of Vietnam) soldiers carried their weapons wherever they went.

We found a little sidewalk restaurant, sat down and enjoyed a delightful meal. After lunch we walked around the village, which didn't take long. Then we headed back to the air field so we wouldn't miss our ride.

We watched the fuel truck head for the compound, then the jeep and its driver went by, we waved and waited. Dower said we should head back up to be sure our gear is secure. We returned and waited. Some of us fell asleep during the wait. It was nearly 1700 hours and we still hadn't heard from anyone at the Da Lat compound. It would be dark in about an hour. We couldn't spend the night there. Sgt. Dower sent Rooster and Chin Lee to the village. He instructed them to hire a wagon, truck or whatever to haul our stuff down and to rent us a couple of rooms in the local inn or whatever they called it.

It was just about dusk when we saw the cart. A large water buffalo pulled it as it approached us on the roadway. Chin Lee hopped off. He and Rooster got us a couple of rooms at the inn. We all looked very cautiously at that buffalo. Our previous relationship with this particular species had been most expensive and embarrassing. The animal snorted and crapped while we loaded the gear on. Then we headed down for the short ride to the village.

We settled into a two story lodge near the airport road for the night. We had only U.S. money and the locals were quite receptive to accepting it. Each room had one single bed with straw mattress and no pillow. We had sleeping bags and the floor would be our mattress. We ate a meal of fried rice with shrimp, ham and chicken (we think) and local rice wine shooters.

Soon after dinner I felt warm all over, rice wine is some mean stuff. Sgt. Dower suggested we post a guard. We decided to put the gear in one room and we would sleep in the other. We cut cards for our shifts. I pulled the king of hearts, high card, first watch, 2200 to 2400 hours. The night passed quietly, sans the occasional barking dog.

U.S. Army Welcome

The next morning, 21 September 1967, a knock on the door startled us. We grabbed our weapons. Dower whispered, "Be calm," faced the door and asked, "Who is it?" "You guys from Nha Trang? We weren't expecting you until tomorrow, got a truck here to carry you to the compound," came the reply from the other side of the door.

We dressed and hauled our stuff to the waiting deuce and a half. Once loaded we climbed on and finally headed to Da Lat. The drive was scenic. Flowering trees lined the roadway along with neat concrete homes with tile roofs. A couple of large Catholic churches took up portions of the cityscape. Cattle pulled carts; beautiful girls walked or rode bicycles. An occasional motor bike passed by.

The U.S. compound was contiguous to the South Vietnamese Military Academy. It was fully enclosed by thick, cement walls and ornate iron pickets. Concertina wire topped the fences. We pulled through two gates made of thick wood and steel bands. Once they closed, they provided a formidable barricade. We hopped off the truck and almost immediately an irate, rotund First Sergeant confronted us. "Fall In, FALL IN!" he shouted. He continued his rant, "Who is in charge of this group and where the Hell have you guy's been?"

Sgt. Dower yelped out, "Me First Sergeant! We stayed overnight in the village below the airstrip."

"And what is the name of the General officer who instructed you to violate your orders to report to this compound upon your arrival?"

"None First Sergeant," Dower responded loudly, "the Army failed to meet us and provide transportation to this location. I made the decision to remain at our general location in case the Army sent a party to look for us."

"Stand at ease troops…and you Sergeant, what's your name?"

"Dower"

"Come with me." The two men disappeared into the HQ building.

After about ten minutes Dower came strolling out and announced sarcastically, "Well we have one new friend here"

The CQ (Charge of Quarters) showed us around. The Mess Hall and PX were down on the end of the complex. The buildings were masonry with wooden floors. Ceiling fans whirled the warm air around. Real beds, singles, lined both walls, eight to a side. It looked like 1940 style lighting with big, white globes and bulbs the size of ostrich eggs. I remarked, "Quite satisfactory, Sarge. Any chance we can live here instead of the mountain? I could really get to like this place."

"Any chance we had of that ended when that fat man, reamed me in his office. As luck will have it, a convoy comes down in the morning; you will be a mountain man by tomorrow nightfall."

Chin Lee was to this point reluctant to be assertive. We were surprised to see him take off for the serving line. Then we saw Anton Brown, our bunker and tent mate from Da Nang. He was now a cook and served up chow here in Da Lat.

"Well," I say, "we have half the New York contingent here, wonder where the other two might be?"

Sergeant Brown joined us as we sat down on the benches along the exterior walls of the Mess Hall. Now we are really shocked. Except for Dower we were all E-2's, as was Brown, just a few short months ago. "How did you get those stripes?" Wilson asked

"Well, when we started getting shipped out of Da Nang to all over the country, I went to Headquarters Company in Saigon. There were so many GI's there with nothing to do, so I volunteered to work as a mess cook. Next thing you know, I'm PFC, next a Corporal and then when they sent me here to run the Mess Hall, made me Buck Sergeant. I am supposed to get Staff in another month."

"You mean in a little over two months you are waiting for E-6?"

"Yep, and the only bad part is I still am getting E-4 pay! They assure me that it will eventually catch up." Brown paused as we let it all sink in. "Where you guy's headed or are you going to be permanent party here?"

Sgt. Dower explained that we were headed for a place called Pre Line Mountain. He asked Brown if he had any information on the mountain.

"Only that they come down here about twice a week for resupply and mail. The place is supposed to be pretty remote, can't tell you anymore about it. There's supposed to be an MP Company and a bunch of mountain people there."

Brown invited us down to the club and offers to buy the first round. Sounded like a terrific plan. We headed the short distance to the club.

I drank one beer and paid for the second round. I left early to write home before I missed the opportunity. I tried to write my wife everyday and occasionally penned a few lines to friends and family too.

The Central Highlands was cooler than the "I Corps" area, (Saigon and the Delta), and reminded me more of North Carolina, red clay and pine trees. The remainder of the afternoon passed. We returned to the club, said our goodbyes with one more round and then returned for our last night in Da Lat. We walked back to the transient barracks and marveled at the tranquility of this place. Other than an occasional Huey or Chinook, we had seen no indication that the people here knew of the conflict within their country. This place seemed like business as usual and normal family life, very strange. It was just about lights out and the loudspeaker played Taps.

We awakened when the billet door opened. The CQ ushered in a Korean officer.

"Hey guy's, this is Lieutenant Huang To Lee. He is going to be with you up at the mountain site." Lieutenant Lee spoke clear English. Lucky for us I guess. We still had a hard time with Wilson's New Yorkese. We introduced ourselves and asked him why he was heading to Pre Line Mountain. His response was pretty direct, "They gave me choice of court martial or remote assignment. I make this choice."

"Welcome to the detail Lieutenant," says Dower, "glad to have you with us."

Lieutenant Lee traveled light. He had only a small duffle bag and his weapon, a unique semi-automatic, wooden stock and receiver and about a nine inch barrel. It appeared to have a banana clip of thirty rounds capacity. His duty uniform was dark camouflage colors, more Delta than Highlands oriented. He had a large knife, similar to a "Bowie" on his right hip and a Colt .45 on his left. This guy had a story to tell. I wanted to hear it, but would have to wait for him to volunteer. We rolled into bed and settled in with our new roomie. Morning reveille came quickly.

I woke up hungry and wasted no time heading for the Mess Hall. I looked over to where our new team member bunked in. No one was there. I wondered if maybe he decided to go back for the court martial after seeing the bunch of warriors he would be with. I got to the serving line and filled up on scrambled eggs (reconstituted, but catsup makes them better), bacon and toast. Hot coffee topped the meal and a cigarette for dessert.

Rooster sat with me. He said, "You know what I seen this morning when I woke up?" I had no idea, so I asked him to tell me. "Lieutenant Lee was out on the porch doing some kind of jujitsu or something. He was chopping and kicking and jumping all over the place, weird man, just weird."

Chin came over about that time and said, "It's Tai Kwon Do, not jujitsu, Koreans do Tai Kwon Do." He explained, "It's a martial arts system used to build agility, reflexes and body toning. This guy is in pretty good shape. And I understand the Koreans use it in their PT program."

"I wonder what he was going to be court martialed for," Wilson asked as he sat down at the table.

Sgt. Dower approached and told us we will likely leave at 1030. By then we should have all our stuff set for loading, weapons loaded but no rounds chambered until we clear the compound. It wouldn't be long now before we arrived at our new home.

Adventures at Pre Line

The supply convoy was not so impressive, two jeeps and four deuce and a half's. There were perhaps fifteen GIs total. The Staff Sergeant in charge pulled up to the CQ office and stepped out. After about five minutes, he came out and the Duty Officer pointed us out. He walked in our direction and Dower reported. "Load your guys on to the second truck Sergeant. You will ride with me on the way back. Be sure that they have everything secure, once we pull out of here we don't stop or slow down until we are entering the site."

The truck driver was tall and skinny, probably in his early twenties maybe. He wore an OD tee shirt, flak jacket and jungle hat. The truck had folded down bench seating. We stowed our stuff on the floor of the bed and folded down the benches, we sat, two to each side. The others in the convoy loaded food, building materials and other equipment. We saw Lieutenant Lee sitting in the other jeep, staring ahead waiting I suppose for us to get underway. Our truck fell in queue with the others.

On the way I noticed we passed three funeral processions. I guess this was one way the locals realized the reality of war, me too. It seemed we traveled seventy miles per hour or faster. It probably wasn't half that fast, the curves and dips just made it seem so. We passed through several narrow cuts that had steeply banked sides twenty or more feet high on either side. I imagined these were prime ambush sites. I looked ahead and on one of those banks saw several uniformed soldiers, waving at us. Since they were waving and not shooting I figured they must be our friends, so we all waved back. Their Highways are comparable to our secondary roads in the States. In comparison to some I have seen here, this could rate as a super highway. At least it was paved and two vehicles could pass safely. The farmers and others we passed on the road yielded to us. Our drivers warned them with the horns. Eventually we turned off the main roadway onto a road we called, Pre Line Road. Vast green areas of tea plants stretched for hundreds of yards on both sides of the road. A small group of hooch's was to our right. About another half mile further we saw a village of about thirty small wood and metal shacks, some with thatch roofing. We slowed when the road inclined and meandered. We saw more little men in dark green uniforms walking the edge heading down to the village or someplace. They carried old U.S. M-1 Carbines and one has a Thompson

I guessed they were mountain people who shared the site we re headed for. We waved and they waved back. How's that for foreign relations? After one last sharp turn, we leveled out on a flattened plateau. Ahead of us was a primitively protected entrance way with large logs from fallen trees, concertina wire. Also, there must have been nearly twenty claymores within ten feet of that opening. The sentry was about five feet tall and a little pudgy. He smiled and waved us through. About a quarter mile down a bumpy dirt road we saw antenna 100-200 feet long and 50 feet high. Two buildings with the words Page Communications painted on the roofs flanked the antennas. A few hundred feet more and we entered another gate. This one was metal and heavily fortified with earthen berm, Connex bunker boxes reinforced with logs and sandbags.

Our complex was inside a 1000 foot perimeter. Five major structures make up the center of the compound: an Ops Bldg., mess hall, Officer and NCO billets and showers.

The communications area was in the Ops center, just outside of a mortar pit dead center and at the highest point on the compound. A 180 foot communications tower was positioned beside the pit. As we drove around I saw more Connex bunkers, a two story tower near what appeared to be the fuel point and motor pool. An ammo storage area was mounded up and covered with a mountain of sand bags off to the left of the Ops building. A bunch of wood, tin and cardboard hovels were at the far corner of the complex, apparently where the mountain people lived.

As I took all of this in I thought, okay then where would we stay? I didn't see any enlisted housing. Then I saw several metal huts, half round in shape and glistening in the sun light. At the very back was a protected bunker open at one end. Several large generators were housed there. I suppose they provided the power for our location. To the left of there were three more of those half round buildings We stopped at the Ops building and jumped off the truck. Sgt. Dower and the Lieutenant had already gone in. We did the "army wait" protocol. Our driver was not too talkative and we wondered if everyone here was as friendly. Then they (Lieutenant and Sgt. Dower) came out with this First Sergeant who must have weighed three hundred pounds, stood probably six foot six and had the longest handlebar mustache I have personally ever seen. He bellowed, "Listen Up! You guys will be staying in the Lima Four Sector, those two hooches over there."

He pointed to the two huts farthest away from everything. What were they trying to tell us? He continued, "This afternoon we moved you a generator and wired you up for power. In the next few days you will construct your own shower and build your own latrine. For the time being use the units back of the compound by the power generation area."

Great! We had to walk a quarter mile to take a dump or shower. He continued, "This is Lieutenant Andrews, the CO. I'm First Sergeant Gross and don't read anything into the name, I am a really nice guy! You are part of a 92 person military police unit. We are part of the 316th Military Police Company and our main base is located in the City of Pleku, South Vietnam.

We have 125 soldiers of this Army of the Republic of Vietnam and their families living here with us. These are Montagnards and most are on their second families, having lost their originals to the Viet Cong. It is important that you do not make any attempt to get involved with these men or their wives. These people are fierce and very protective. They are territorial and have a very high disdain for the Vietnamese national population."

Lieutenant Andrews spoke next. His introduction was more to let us know we would be patrolling the roadway into Da Lat. We would alternate with the Montagnards, and occasionally patrol together. We would pull guard details at the Lima side bunkers and join patrol duties starting in a few days. Normally we would do sweeps about six miles to the Alpha Checkpoint and return. Two per day, our contemporaries (the mountain guys) do one and we, the other. The Montagnards covered nighttime activities. We had three interpreters assigned here one went with us on each patrol.

Lieutenant Andrews also explained that local villagers would cook and maintain the compound. The Village Chief was our official "hygienic maintenance technician". In other words, he daily burns what we dropped in the "one holers".

Chow was served from 0600 until 2000 hours coffee was available 24/7. There was a Club where we could get essentials, beer, sodas, candy and cigarettes. We were advised to always have our weapons and flak jackets near; steel pots (helmets) were suggested we could however wear the jungle hat, a camouflage olive green head cover.

We jumped back on the truck and rode over to our new home. Both of the hooch's had one door on either end and about halfway up the arched roof were fiberglass panels that let in light. A large fan was mounted through the wall above the entrance door. A louvered vent on the other end brought fresh air in. There were twenty bunks, single file, ten to a side, double door wall lockers between each set of bunks and a foot locker at the base of the bunk.

"Gentlemen, this is home, said Sergeant Dower, "I will be bunking at the NCO billet, get settled and then we will familiarize ourselves with this compound, especially the posts and bunker provisions."

I set about making my stuff look orderly in the locker. No hangers so everything had to go flat on the shelves. I set my weapon in the open area and my flak jacket too. I stacked my footlocker on top of the one provided. I didn't have enough stuff to fill both of them, five sets of fatigues, 12 pairs of socks, eight sets of underwear and t-shirts all in the fashionable OD green color as were my eight towels and six wash cloths. I also had two pairs of jungle boots, steel pot, flack jacket, field jacket, goose down sleeping bag and one OD wool blanket.

After I unpacked, I loaded empty clips for my weapon. Though there were larger clips, these held thirty cartridges and we taped them together so we could snap them out and turn them over to reinsert into the receiver pretty fast. I hadn't had to do it yet, so I relied on what I had been told by others.

Since my arrival in Da Nang, with the exception of storming the beach, we have been provided with ample ammunition.

After writing just two sentences of a letter to my wife, Dower comes in with Lieutenant Lee. He said, "Okay fellas, let's take a walk and see just what this place has."

Taking the Tour

We grabbed our weapons, flak jackets and soft hats. Dower lead off. Lieutenant Lee followed and the four of us trailed him. We stopped at the bunker nearest our hooch.

It was a Connex metal container, modified by cutting out an opening from side to side about thirty inches wide and six inches down from the top. A huge battery sat on the left of a shelf in the center. Twelve wires tightened down on the negative post. On the right were twelve more wires, insulation stripped back about two inches and the ends of each were tightly twisted. A drawing on the shelf marked the location of the claymores. Each wire had a number that corresponded with the claymore location. The locations of Bouncing Betty trip mines were also marked and strung with trip wire interspersed among the claymores.

The mines popped up when the wire snapped which caused them to detonate about two or three feet off the ground. It could take the legs off at the knees and leave hundreds of pieces of shrapnel. Two rows of concertina wire and "betty's" between each row before the claymores are reached on the third perimeter. Each of the bunkers was similarly equipped. They were about 60 yards apart. Our group would man the two nearest our hooch at intervals of two hours on and two hours off, every night. These bunkers also had an M-60 set up and belted for immediate use as well as an M-79 grenade launcher.

We continued along the "berm" (earthen barrier which the bunkers were inserted in). They served as the last barrier between us and them in the event of an attack. The pathway led down an incline where we saw the two story tower that was built prior to our arrival at this site. It was pockmarked from small arms and fragmentation grenade strikes.

The fuel depot was made up of three separate sand bagged rectangles. Each had several fifty five gallon barrels stored by the type of fuel they held (Diesel, Mo-gas or kerosene). They were equipped with hand operated fuel pumps, the kind you cranked. Of course there were more diesel fuel barrels as it powered the generators and about half the motorized vehicles on this compound. The jeeps and three quarter tons use mo-gas, no idea what the kerosene was for, we would find that out later. There were no bunkers at this corner, though there was one adjacent to each side, within thirty feet. I guessed the potential for fire or explosion made it imprudent to put people nearer than that.

We continued our tour and walked past the generator bunkers, the billets and other bunkers, each set up just like ours. I wondered if this was standard army protocol or just lack of imagination. The berm was about twenty feet thick and ten feet high which must have been a challenge for the engineer company that built them. We entered the Montagnards housing area. There we saw young women in their teens or early twenties at best, naked from the waist up. Most had babies tied to them in some type of cloth sling. The children all seemed to be under three or four years old. The kids were naked. They don't seem to care I guess it's the way they lived. Shortly after our arrival, all the females were issued their own t-shirts in olive drab green.

They lived in shacks no more than six or seven feet square. The shacks were made of cardboard, with wooden slats and tin roofs. "Hey look there," said Rooster. We saw an old man pulling barrels out of the trap doors in the back of the latrines. Some barrels were burning with a bright orange flame, others were just smoldering.

"That must be Chief Poopie," Sgt. Dower said. We all chuckled, "Chief Poopie". What

a name to get hooked with". We now realized what the kerosene is for - feces fuel. As we walked by Chief Poopie nodded and smiled.

Wilson exclaimed, "Hey his whole mouth is purple, what the hell is that?"

Lieutenant Lee, who had not spoken a word up to this time, explained, "These people chew on Beetlenut. I understand it has a slight narcotic effect. Both women and men villagers use it. It does discolor the gums and teeth with that purplish tint."

We neared the center area. Dower wanted us to see the Commo room and mortar pit. The pit was only about ten feet in diameter and they had storage for the mortar shells all around its perimeter. They can fire high explosives, white phosphorous, and illumination flares from this pit. The communications shed was really the box off a deuce and a half. It had been sand bagged in and the commo tower was right beside it. Directional antennas on the top boosted signal to relay radio traffic, both locally and to remote sights.

We saw some young Vietnamese girls at work around the compound. They were apparently the workers from the village below, hired to keep the place clean. First Sergeant

Gross came out to meet us. He said something to one of the interpreters and shouted to a young girl who was sweeping the steps to the Commo box. She came over, obviously nervous and shy as we inspected her.

First Sergeant Gross said, "This little gal will be your housekeeper. She will clean your hooch, make your beds, polish your boots and launder you clothes. She will not be your sex partner and if any of you attempt to seduce or harm her you will feel the total wrath of this mans Army! Am I clear on this?" "Yes First Sergeant," we responded in unison.

He continues, "By the way you have to pay her for the work. She has agreed to ten dollars a month, and your first month is due now." That was the quickest ten bucks I'd spent in a while.

We paid and continued our tour. We headed for the entrance to the main compound and checked out the fortifications there. The Montagnards were assigned daytime guard duties at most of the posts. They were posted at the entry to the compound and conducted security checks of the villagers as they came and went. The history of this relationship, as I have heard it, is that the Vietnamese used to treat the Montagnards very harshly. Rumor had it that they actually hunted and killed them for sport. They regarded them as animals and the mountain people developed a very strong dislike for the "civilized people" of South Vietnam. They were commanded by an RVN Captain and a Montagnard NCO. According to our interpreter, those two men were not on the best terms either.

Just outside the gate was a six foot by four foot cage covered with concertina. "What is that for?" I asked. The interpreter answered, "The cage is for soldiers (Montagnard) who break the rules. They are placed there for three days and only get water during that time." Apparently it was effective, as it was empty that day.

"What about the Page contractors?" someone inquired. They work outside the compound, where do they go if we get hit?"

There is an underground tunnel that carries them back into the main compound. It is rigged with explosive charges and can be blown if it is compromised. Man, pretty complex set up isn't it.

Sgt. Dower suggested we head for the mess hall. Guard duty will begin tonight and he suggests we get what rest we can. An NCO will wake each crew for shifts every two hours. The mess hall was somewhat Spartan in accommodations, just benches and tables. They cooked on an open covered porch and we served ourselves. Regular U.S. grub, fixed the army way and seasoned to your taste with salt, pepper, and whatever else helped. This was actually our first opportunity to get a look at the rest of the GI's here. They appeared to be normal, loud, and cheerful to a point and welcomed us to the mountain.

After chow we reported to the Ops building and were briefed on activity in the area and current intelligence. Then we went back and waited for our time in the bunkers. Chin and I were in the second Lima bunker. He took the first shift and we swapped off through the night.

The next morning we grabbed some chow and headed for the bunks. While Rooster was dropping his skivvies prepared to head for a shower, we heard the shriek. We turned to see our housekeeper reporting for duty. Lucky for us Rooster pulled up fast. The noise scared him. We all proceeded to the showers with clean items and toilet kits, wearing fatigue pants and flip flops. We passed the latrines and saw Chief Poopie doing his duty.

He bowed and smiled the purple grin. We responded with our own grins and bows. It took about half an hour to get inside the showers because of the line. We got five minutes to shower, five to shave and then you were out. By the time we get back to the hooch, two things have happened. One, our feet were filthy and two, the housekeeper had completed cleaning up. She gathered our dirty stuff and put marks on them to tell her whose stuff was whose. Our names were already on everything, guess she just needs to put it in Vietnamese. Wilson asked her what her name was. She just looked at him and said, "Com biet."

I guess that meant, "No speakie English." He then pointed to himself and said, "Wilson." She smiled and repeated, "Wilson. One by one we told her our names and she repeated each. She then pointed to herself and said "May Lene" or something like that. From that point on we called her May. She nodded and we have established our line of communications. I suggested we have one of the interpreters come over so we know what she wanted us to do and provide, like soap powder, shoe polish and the like. Day one with our housekeeper seemed to have gone well.

After four days on site a convoy was heading down to the Da Lat compound. There was no rhyme or reason for the frequency of our trips down site. The Montagnards filed out on their first sweep. Instead of taking the road, they cut through the hillside and the tea plantation to the main highway. They patrolled the roadway into Da Lat and returned a few hours later via the, Pre Line Road. The convoy passed them on both legs of the trip. By the time they got to the main road our trucks would be on the way down. None of us went on this trip. We were still too green to travel with "veteran" soldiers. I just hoped we got some mail. It had been a while since any of us got letters from home. I was concerned about how the new baby and my wife were holding up.

Sergeant Dower interrupted my thoughts by telling us it was time to build our own latrine and showers. We went to the motor pool area to get a three quarter ton and load up some materials we need. I signed out the truck and met the rest of them back by the generator site where all the lumber and construction stuff was kept. We grabbed some plywood, two by fours, hammers and saws. Sergeant Dower sent Chin to the Ops building to get an extension cord and power saw.

We went to work. Our housekeeper May experienced much humor in our efforts at the construction effort. She showed her amusement with constant giggles as she passed by on her travels. Since we didn't think about measuring anything we just used the whole boards as we built the frame work. Sgt. Dower soon corrected this by making a yard stick out of a two by four. He said that a hand saw was exactly thirty six inches long so he laid the saw on a board and marked it on the end and sawed it off. Then he proceeded to cut off all the extra pieces that projected from places he did not want them to be. By the afternoon we had some form of an out house framed up. Good work. Now let's go eat some lunch.

Ambush

While crossing to the mess hall we heard small arms fire and short bursts from semi automatic weapons. We concluded quickly that since the Montagnards didn't have large caliber semi weapons, either the returning convoy got hit or the Montagnards had been ambushed. We automatically headed for our bunkers and waited.

Sgt. Dower headed for the Ops office to see if he could find out what is happening. He returned rather quickly and advised us that the Montagnards were ambushed on the roadway to the site. It was a typical hit fast, hard and then run away attack. Several were wounded and one KIA (Killed in Action). The First Sgt. was taking a squad down with a couple trucks to bring up the injured. The medic would stabilize the wounded that needed evacuation so they could be taken down to Da Lat by Medivac chopper. Doc asked if anybody was first aid trained he could use some help. I said I would give him a hand my firefighter training course included first aid.

The aid station was a small shack about the size of the latrine we were building. There was enough room to put two narrow exam tables and you stood between the two to work. I got there just in time to help him set up IV's and the basic surgical tools he would need: a scalpel, several types of tong like tweezer looking items, scissors, gauze pads and he had loaded some syringes with whatever. Though there was a single bulb fixture inside, the open door was the best light. Doc told me to lay a couple of ponchos outside the doorway so that he could triage the most critical to least critical. We had no idea how many would be brought up. He was certain that the first patients would be here in minutes.

The medic shack was alongside the Ops building at the edge of the mortar pit and the vehicles could get right up close to it. The truck slid in with dust flying. The first casualty looked so small, almost swallowed up by the litter itself. I looked in disbelief. His left leg from the knee down was standing straight up on the litter, still in his boot, but not attached to his body. Doc had them bring him directly into the shack and laid him on the table.

"Start an IV," he said. I pulled the tri pod stand closer, took the scissors and started cutting the shirt off his arm. I wrapped a latex band tightly above his elbow and the vein popped right up. I inserted the IV needle and started a slow drip.

Doc said, "Open it about three quarters, we need to get fluids in him." He talked casually as he stitched some large wounds. He related how the body adapted to trauma in unique ways. The amputated stump was not bleeding profusely due to the constriction of the circulatory system. It was the body's effort to survive. The calm voice switched quickly, "Change bags!"

I put another bag on the stand and inserted the tube into the IV needle, setting it at three quarter flow. By this time another victim was on the adjacent table. This fellow looked like he had measles or severe acne. Doc explained that he was peppered with shrapnel from a frag grenade. This guy needed an IV also. He was more likely to go into shock than the guy we were working on. I stepped over to the table, grabbed another IV kit, set it up and banded this fellows arm for sticking. No vein, Doc," I say. Reach over and try the other arm, he says. Okay got a vein, stick him and set it to flow. Make it about half flow. Keep an eye on these two while I check to see what's outside."

He dressed one with a clean wound through the forearm. The fourth guy had some frag wounds and a gash on the forehead. Doc put three stitches in and walked back into the shack.

Doc came over to me, grabbed my arm and pulled me to the doorway. I stumbled at the door and fainted flat out like a sissy. The combination of warm temperatures, close proximity and the coppery smell of blood had abruptly ended my medical career. I regained my wits only to hear the guffaws of my comrades in arms as they relived my passing out and sprawling face down into the mortar pit. Thankfully, nothing was broken and only my pride was injured. I thanked them for their concern and excused myself to wash up.

The convoy was directed to stay in Da Lat overnight. The Medivac took all the injured down for further attention. It was already time for supper. At the mess hall Doc came over to thank me for my help. He told me many men faint. He admitted that even he still got woozy on occasion. He asked where I learned first aid from. I told him about my fireman experiences. We chatted for a while. He thanked me again and said he might have to call on me for help in the future teasing that the next time he will put a fan in the shack just for me.

"Okay Doc, do that and maybe I can stay conscious longer," I laughed.

Doc left and I resumed eating. The food seemed to taste better than usual, possibly because we missed lunch. Sgt.

Dower told us about the attack. "From what I understand, they were hit with crossfire of thirty caliber and fragmentation grenades as they walked through a steeply banked gully along the roadway. No good cover for them so they had to crouch and fire. They are not really happy with their leadership and felt that the RVN Captain had not supported them with better weapons and equipment. They blamed him for the injuries and the loss of one of their fighters. The mood in the Montagnards camp is very somber and one of mourning."

We settled in for another night of heightened alert. It was going to be a long one, knowing that the enemy forces were so close to us today. Come on dawn, a warm breakfast and a little sleep will make our anxieties diminish. I wanted to work on the latrine the next day. It would occupy our minds and bodies.

Several days later, the anxiety had pretty much passed. The hostility between the mountain folks and their RVN Captain continued to increase. They were reluctant to obey his orders and according to our interpreter were ready to rebel against him. I was impressed that he had not called for additional RVN soldiers to protect him.

We worked on our own shower, making it an engineering marvel. Three sections of antenna tower on top of which we laid two by fours covered with plywood. We placed a two hundred gallon water tank on the makeshift deck after disassembling it from a trailer. A piece of rubber hose ran from the spigot on the tank and attached to a length of copper that eventually was coiled into the bottom of a fifty five gallon barrel. This barrel was mounted on top of the first section of antenna tower. A five gallon Jerry can of diesel fuel fed by a drip tube into the larger barrel. The copper coil ran out of the fifty five gallon barrel into the shower stall at the antenna tower base. Just a couple pieces of plywood covered the wooden slatted base for drainage. We drove nails into a canteen cup to make a shower head. Water flowed into the hose, then the coil, the diesel fuel dropped into the barrel and burned. As the fuel heated the water and coil warm water ran into the shower cup head. Taking a shower was a two person operation, but worked. The final task of filling the tank required a bucket brigade and took several hours to accomplish.

Lieutenant Lee moved into our hooch. He pretty much stayed to himself, though he worked on the shower with us. Every morning and afternoon he did Tai Kwan Do for about forty five minutes each time. Lieutenant Lee wanted to go on patrols with the GI units, obviously bored with his inactivity and lack of involvement.

Rooster and I went out with the next convoy to Da Lat. Sarge wanted us to rotate on trips so we wouldn't go crazy being stuck there all the time. Rooster was the bachelor of our group. He wanted to find a girlfriend for an hour. The guys told him they cost five bucks for a night and two bucks for an hour.

Wilson and Chin shouted into the doorway of the hooch, "Let's eat!"

We headed out for the mess hall, food on our minds. It was Thanksgiving Day, 1967, and we had a feast with turkey and all the good stuff that goes with it. I thought about home and what I would be doing if I was there instead of here. My family would get together, at least her family. Mine is so strung out and fragmented.

Back at the hooch, I lay on the bunk and thinking, why did I eat so much? We were all relaxing for the afternoon, until duty called. At about 0200 hours, 25 November 1967, a bouncing Betty tripped at Lima 4 sector near the motor refueling point where the two story tower stood. They opened up with the M-60 and Lima 3 supported with an M-79 HE round and M-16 fire.

Eventually the gunfire ended. The rest of us were in bunkers and straining to see. The mortar crew popped a flare and I heard that "flooping" sound. In a few pounding heart beats the igniter lit and the reddish white light illuminated the sky. We didn't see anything with the light of the flare. Who or whatever it was left or dug a very deep hole very quickly. As that flare sank below the berm another was fired, and soon it too wafted down, nothing evident. We stood down to normal posting and the night passed into morning daylight without incident. That morning we sent a detail out to check the location and replace the tripped Betty. I was pleased I didn't get that detail – I didn't want to think about trip mines, they are very unforgiving.

A few days later we finished the shower and began to experiment in regulation of the fuel into the barrel. Too much fuel made it too hot, and conversely too little left us chilled. That task resolved, we returned to routine.

It's 28, November 1967, and Wilson is on the patrol duty list. As he was getting his stuff in order we learned that First Sergeant Gross was taking the detail today. They usually met by the concrete tower near Lima 4 and moved out from there. Wilson was uncertain what to expect.

"Hey, you form up like he tells you, cover your spot and react to whatever happens, not to worry," I told him. My encouraging commentary did not comfort him. He trudged off and we resumed our activities.

Chief Poopie was on the job that day burning out our latrine barrels. He usually finished at 1400 hours and the baby sans and mama sans worked until 1700 hours. While he waited for them to finish, he sat between the hooch's, lit his opium pipe and nodded off on his drug induced slumber. The younger girls came by and collected him in a wheel barrow and wheeled him home to the village.

Between the beetle nut, opium, hash and other available drugs that are openly used, I wonder if some of our people weren't using. Maybe they were. Some of the folks on this site were really unusual in their actions and interactions with others.

Wilson returned to the hooch. For sure they haven't gone out and returned in this short a time. "You guys will never believe this," Wilson exclaimed. "Some Brother, they call him Cincinnati, was down at the briefing playing around with his M-79. He was trying to spin it like a six gun and didn't set the safety. He popped a round right into the middle of the group. We were assholes and elbows trying to get away. The round landed right on top of First Sergeant Gross' topo map. He had the guy in the Ops building and he was not a happy man. The good news is no patrol duty today for this detail. The troops for tomorrow's sweep are taking it over."

Convoy to Da Lat

Sergeant Dower came in the hooch and said "Rooster, you and Simpson will be going on the convoy to Da Lat tomorrow."

The anticipation of going down site made the night go by quickly. I looked forward to a change of scenery, people and buildings sounded exciting. We slept a while, ate lunch and slept some more. Before long it was time to eat dinner and prepare for guard duty. We napped until we were called and the night passed. Another dawn rose up and lit the horizon with sunlight. I rushed over to eat and get washed up before the convoy pulled out.

We climbed up on the deuce and a half and settled on opposite sides of the bed. A metal tread plate fastened to the side rails and sand bags covered the floor. They were our only protection from roadside grenades, booby traps and small arms fire. I was determined that if anyone shot me I would get a flesh wound in the buttocks because that was the only part of me I couldn't hide.

On the convoy, was the lead jeep, three deuces and the second jeep followed behind us. Out the gate and down the Pre Line Road we went. Dust flew as we traveled fast through the winding and narrow passageway. We came to the intersection with the main roadway and made a hard left turn.

Now that we were on paved roadway we picked up speed. It appeared our driver knew the drill: go fast, honk your horn and don't slow down. We passed ox carts, laden with firewood, caged poultry for the market, wood for building and all sorts of local merchandise. I also noted that we passed by a couple of carts hauling caskets. This area was experiencing considerable loss of local soldiers. As we neared the city we saw more traffic, bicycles, mopeds and scooters, only a few cars and of course buses. I knew Wilson is out on the patrol today. I wondered if he spotted us.

We arrived in the city and approached a unique large stone building. It was a Catholic church with a steeple about three stories high and several bells on the top. A burial ceremony was going on with only a couple of workers and the casket. How sad that there were none to mourn or pay tribute, I thought.

We entered the compound and proceeded to the storehouse to load up our supplies for the trip back. We packed food and perishable goods in OD coolers and loaded them on the first truck. The second truck got the fuels. Mail and beer, soda, candy and cigarettes and stuff like that went on the third truck.

We loaded and were ready to return within an hour. The convoy leader's driver told us not to wander too far. It was almost noon and Rooster was not going to have his way with any women today. It seemed I wouldn't be calling home either. Convoy's always altered their schedule. This kept the potential for ambush less predictable. If the VC, don't know when you plan on moving out, they have less opportunity to set up a site along the way. Too mush risk of them getting caught by our patrols as they secure the roadway. We opted to go to the mess hall and grab some chow. Perhaps Brown would be there and we could shoot the breeze with him. We didn't see Brown. We figured he had the day off. We dumped our trays and headed outside for a smoke.

We saw a large group of Vietnamese folks, mostly young women and a few old men, at the exchange office. I asked one of the guys who passed us what was happening. He said that this is currency exchange day and a number of the young female business women were trying to exchange Monopoly money for Piasters.

The local script changed frequently (to discourage counterfeiting). Citizens were permitted to exchange the script for Vietnamese currency once a month. The Red Cross girls (Donut Dollies) visited this site and passed out games. The GI's thought it would be funny, and more economical, to pay the local prostitutes with $500 and $1000 Monopoly bills. When the Quartermaster did not honor the funny money, the girls were somewhat upset. They gave up a lot of free booty this past month. I told Rooster, I think we just witnessed the last of the five dollar all-nighter in Da Lat. These ladies have a lot of losses to recoup.

First Sergeant Gross yelled, "Load Up!"! We jumped back on the deuce and a half for the trip back up the mountain. We were out of the city in no time, cruising the highway. We passed a of couple trucks with RVN soldiers on board, headed for the city I guess. Not much traffic out here now, I didn't know if I should be concerned about that. The locals seemed to know the VC plans and made themselves scarce if activity was expected. We continued on our route and almost halfway to our cutoff, we started slowing down.

Both Rooster and I looked around to see what might be going on. There was some type of blockage on the road ahead; light smoke wafted up from that point. First Sergeant Gross halted the convoy. We dismounted and took cover on opposite sides of the roadway. The driver slid under the truck with his weapon ready. The two guys on the lead truck slowly advanced to the position and reported back to the First Sergeant's jeep. An ox cart had struck a booby trap and the carts contents were all over the road. The Ox was dead and the driver was severely injured. There were no signs of any Victor Charlie. The poor slob just happened to go off the roadway at the wrong spot. First Sergeant Gross radioed for medical assistance and sent the guys back with a first aid kit. The First Sergeant yelled my name and told me to go up there and see if I can do anything for the injured guy.

I wanted to say, hey, I passed out last time, what should I do this time, throw up on the victim? All that came out of my mouth was, "Yes First Sergeant!" I headed forward and encountered the old guy or maybe he just looked old, it was hard to guess. Chances were good that he was too old to serve the RVN's and not worthy to be a VC. His injuries were mostly puncture wounds with one deep laceration on his left hip and buttock. I didn't think he was going to expire on us any time soon. I offered him some water and put some compress dressings on his more prominent injuries. Then we just waited to see who or what was coming to pick him up.

It wasn't hard for the Medivac to find us on the only road to Da Lat, that's for certain. We soon heard the familiar Huey sounds. Soon it landed and loaded the victim. Once they were airborne we loaded back onto the trucks and resumed our return trip to the mountain. As I moved to the front of the truck bed I heard this whining sound. Rooster tucked something into his flak jacket. When I asked what he was hiding, he pulled out a little brown furry puppy and said, "I figured if I couldn't have a woman the dog would have to do."

I chose not to argue with his analogy or comparison. He was satisfied and that was the important thing.

We returned to the mountain. After Wilson met Rooster's pet he decided he didn't want to be pet-less either. He came in from his next patrol, 30 November 1967, dragging a most uncooperative monkey (genus unknown), at the end of a tethering cord made from old parachute cords braided together. We instructed him that he could not keep that thing in the hooch.

"Yeah, I know. I'm gonna tie him to the cable reel, the pipe will be strong enough to hold the tether and one of the Montagnards is trading me some light control wire for a pack of smokes."

"How are you going to keep him from chewing through the cord? And how are you going to get the wire around his neck?"

"I got a leather watch band for a collar and a swivel off the M-1 Carbine I picked up today."

"Lot's of luck my friend," I said.

Wilson ignored me, "Oh and fellas, there is one thing you need to know, they were going to eat Freddie, so I traded him for some canned meat. I saw some tins of corned beef in the storeroom at the mess hall. I told them I would give them two tins for him."

I couldn't control my anger at that point. He was going to steal food from us for a monkey, "For Pete's sake Wilson that is theft, a court martial offense, all for a flea bitten monkey!"

Sergeant Dower walked into the hooch at this point and said, "No animals will sleep in this hooch!" Wilson tied Freddie the monkey to the cable reel. Rooster put Shep the dog into a foot locker turned on its side with a towel for bedding. It was beginning to look like we are setting up a petting zoo. I had first shift on guard duty tonight so I headed for the mess hall.

Back at the hooch, the monkey struggled to get free. Rooster brought some scraps for the dog and he gobbled them up in seconds. Then he looks at Rooster as if to say, "Is that all?" We get set for guard duty and I head for Lima 2. The night was quiet and we started a new week at Pre Line Mountain.

The next morning we heard a Huey coming in as we ate breakfast. The Red Cross sent a couple of the Donut Dollies to entertain us. We stayed in the mess hall and enjoyed their efforts to provide us with some fun and games. First thing we noticed was there were no Monopoly games. We got Chinese checkers, playing cards, cribbage boards, bingo and dominos. After a few minutes of hearing a female voice I got homesick for my wife. I went back to the hooch, took a shower and crashed for a nap.

I heard the visiting young ladies board the chopper and depart, another job well done.

The next few days passed by rather pleasantly with entertainment from our petting zoo to keep us amused. Shep the wonder dog gained weight and got even more brazen. He begged for food. Wilson did extra duty on his Article Fifteen for appropriating unauthorized food products. Freddie threw feces and tried to bite anyone who got close.

One day after May finished cleaning the hooch I heard her yelling at Rooster about Shep. Apparently, the little fellow, by now was rather portly, has decided to pee on all the bunks and footlockers marking his territory. He seemed to enjoy doing this just after she completed mopping the floors. Between Freddie the poop thrower and Shep the pee doggie, May's patience was surely tried.

Some time later Freddie started a new game. He tried to get you close enough to jump on you. He then tried to sit on your head and crap on you. He graduated to more gross activity as time went by. Wilson was his only friend and probably the only reason he was still alive.

May did not approve of the monkey and let us know in her own way by pointing at it and yelling something. I guessed Vietnamese bad words. What an interesting place to be at this point in life.

Realities of Warfare

I headed to the Operations Office for mail call. I was hoping to get some letters from back home. Yep, that's me, and me again and me again. Wow, seventeen letters. Wonder who sent them. Sherrie, my wife, sent eleven. Three were from my best friend Ron. The others were from a few of my firehouse coworkers. Now I needed to get them in proper chronological order right so I don't read them out of sequence. It was sort of like getting a bunch of presents for your birthday, the excitement and all.

Mail call was when I really missed being at home with family and friends. Sherrie sent a black and white picture. She would deliver in March. I wondered if that little bundle would be a boy or a girl. Her letters explained that she was still working and the baby was still kicking. She was trying to walk between two parked cars the other day and her stomach got wedged between the cars. She had to back out and go around. She didn't look that big! She sure looked beautiful though. It was wonderful to hear about things at home.

Ron joined the National Guard to keep from being drafted. Earlier he tried to convince me to go in with him. I said, "No way! I have a draft deferred job." Ha! He was now home with his wife and look at me. He chose to remind me of this in his letters on a regular basis. He and his wife were taking good care of Sherrie for me. I was thankful for such good friends. True, her family was there, but she needed some other contacts as well. For sure my side of the family is not going to be of any use, they can't help themselves at this point. Still I love them and miss them too.

My contemplative mood from mail call was interrupted by the sound of small arms, machine gun fire and a Huey gunship passing nearby. We instinctively grabbed our weapons, flak jackets and steel pots and headed for the bunkers. The Huey was called to a suspect area and has taken some ground fire and they are returning same. We saw the ship bank and come back toward the spot again, fire several high explosive rockets, and then climbed up from their attack. We waited until the white smoke dissipated. Other than the Huey's engine noise, we heard nothing. They held for a few minutes and then proceeded in the direction of the target. An unexpected smoke trail from the wooded area was followed by the explosive sound of the Huey being hit. It turned in tight circles and spun down to the tree line in a smoky spiral. It was less than two hundred yards from our position, just down the berm and into the trees. We heard the thud as it hit the ground. We didn't see any fire and prayed they survived the crash.

We headed to the motor pool for a rescue convoy. We couldn't go over the berm. Our own defenses of concertina and bouncing Betties would make it almost impossible to get there quickly. Our only option was trucking to the area and getting to the crash site on foot. First Sgt Gross took the lead jeep and seven of us followed in two deuce and a half's. We swung out of the compound.

It's hard to describe the effort required to stay in a violently bouncing truck while trying to keep a sharp eye on the surrounding terrain and holding your weapon in a defensive position all at one time. We slid to a stop, dismounted and slipped into patrol staging. It only took ten minutes to reach the downed Huey. We arrived and saw no sign of the crew. The M-60's have been stripped off. The PRC-45 crackled and we got traffic from aerial support. Two more gun ships near us and should be in view in seconds. The Huey pilot will look for bodies.

Our group was directed to stay with the chopper until a Chinook or sky crane retrieves it. We stayed posted until that operation was been completed. After what seemed an hour the Huey reported no sign of activity. Soon after I heard the Chinook and its distinctive motor/rotor sounds overhead. It lowered the load master and related strapping. In twenty minutes he had it ready on the tow line and they were hoisting it up. We picked up the door and other parts that had been dislodged by the crash and carried them out and back to the compound. While we were occupied with this activity the Montagnards sent out two scouting patrols to fan left and right of the downed bird. They were unable to locate anyone during their sweep activity.

Once we returned to the compound we were advised that two LRRP (Long Range Recon. Patrol) teams were being airlifted here to initiate efforts for location and recovery of the four missing crewmen. We would support and assist them as required. We were put on heightened alert. Throughout the night our mortar team periodically fired off Illumination flares for visibility and as a signaling method for the chopper crew, in the event they are lost and or disoriented.

That night meant cat naps and not a lot of restful sleep. Our normal anxiety was increased by concern for the missing air crew. Finally, dawn arrived and we stood down to normal security. After a trip to the mess hall with heavy doses of coffee, we loaded up to return to the crash site. The two

LRRP teams started out and advised they would contact us if they find a trail or anything suspicious. So we set up a perimeter and waited for the call.

At lunch time they still hadn't located anyone. We ate C rations. I got spaghetti in an OD can and warmed it with a little chunk of C-4. The stuff burns like Sterno and is pretty safe unless you use detonator cord or caps with it. Anyway, there were two wafer cookies, some peanut butter with the oil floating on top and a can of peaches. I also got the vintage pack of four Lucky Strike cigarettes and matches.

After the meal we waited some more. I was getting a bit lazy when two of the team members came out of the tree line. They had located the crew and all four were dead. He requested litters and some ponchos so we might carry them out and asked for an evacuation chopper to take the teams and the four KIAs back to their site in Plekiu. Wilson and I carried the folded litters and several of the guys gave up their ponchos. We headed into the trees and wove our way to a depression that had been camouflaged to look like natural cover. The other LRRP team members were waiting for us, took the litters and entered the area to retrieve the bodies.

The four were mutilated and tortured. The two Warrant Officers were disemboweled. The two enlisted had their throats cut. They were burned, slashed and penetrated prior to being slaughtered. They died within the last few hours as there was no sign of rigor mortis. What happened to the Geneva Convention rules we heard so much about? This type of inhumane treatment wasn't supposed to happen in a "civilized" war. Everyone had a litter position and we gently laid each body on the litter and covered them with ponchos. I remember the airfield at Da Nang and the towering stacks of coffins. These guys would be going home in ones like those.

I shivered and felt the chill crawl down my back. This could be me.

It wasn't a long trek to the LZ. The Chinook settled there with the ramp down. We loaded the bodies on board. We saluted and shook hands. Hardly any words were spoken. They lifted off.

"May they rest in peace and may we avenge their sacrifice," First Sgt was visibly angry and red faced. He got in his jeep and we all loaded up for the trip back to the compound. After chow that evening the CO held a memorial service for the men. We didn't know their names, where they were from, or if they had families that would miss them. We did know they had made the supreme sacrifice and they died while protecting us. God bless them and their families.

Also, the M-60's were not recovered from the plane. That meant they would be used against us, somewhere, at sometime.

Tai Kwon Do Lessons

Our Korean partner, Lieutenant Lee offered to teach us Tai Kwon Do. He was an eighth degree black belt. Taw Kwon Do was a major part of the South Korean Army physical training regime. I had always been intrigued by these martial arts moves. Four of us from our hooch and two others from Henry sector agreed to lessons. The Henry sector guys included Shockly, a surfer wannabe from California and Greene, a red haired nineteen year old Jewish kid from Florida. During our first meeting 2 December 1967, Lieutenant Lee began by telling us we would be barefoot during the training. So we cleaned the floor areas of glass, nails and other sharp objects.

He made arrangements with his friend in Nha Trang to purchase the traditional apparel for novice students. It cost us five dollars. Fair enough, we paid him the five bucks and started setting the place up for instruction. I helped move and stack all the stuff at one end of the building and then walked outside to witness Freddie the monkey throwing small rocks and dirt clods at anyone who walked near him. He also screeched loudly, all the while trying to get loose from his bindings.

We had our first Tai kwon do lesson wearing our white martial arts outfits, with white belt (signifying rookie status). We lined up in a row and Lieutenant Lee started us off with limbering up movements. We twisted and contorted, raising our legs up as far as we could. After only a few minutes of these exercises I was drenched in sweat and wondered if I really wanted this. We spent a few minutes on controlled breathing exercises. Lieutenant Lee said this would give us greater endurance to perform at maximum levels. Then he told us to stand alongside the hooch walls so he could demonstrate the first series of patterns we would learn.

We learned a series of movements that enabled us to use all our body parts in the defensive attitudes Tai kwon do was designed for. "Wait, defense, we thought this was kick ass stuff," Wilson said.

"That is where you are wrong," Lieutenant Lee answered, "This is strictly designed as a way to protect yourself from an aggressor. True some of the responses will be to inflict injury in a disabling form, but only to enable you to escape or incapacitate your enemy. Ninety percent of the actions you will learn from me are for deflection of blows to protect vital parts of your body."

He started in the normal position of Humility and Respect. From there he completed a series of moves in ninety degree repetition. He finished four separate patterns in precise and swift flowing movements. He was flawless and left me wondering if I would ever make it look that smooth? He then instructed us to repeat the limbering exercises. We did before we were dismissed and told to do them twice daily until our next session two days from today. Oh and cleanliness was required. He expected our uniforms to be clean and correctly worn for each session.

After the lesson, we rushed for a shower and broke for lunch. Afterwards I needed a nap. It must have been all that exercise. At about 1500 hours I woke up to the sound of rain. The Central Highlands were not like the more tropical areas (the Delta, etc.). It rained hard but not daily and most of the time just for a few hours. I started writing a letter while the others cleaned weapons, read or just lay around. We still had a couple hours until chow.

The rain stopped as Sgt. Dower came in and announced, "There's a convoy to Da Lat tomorrow afternoon. You will stay overnight and return the following day. Chin, you and Wilson have first dibs on this one."

Chin Lee said he was not interested so Rooster, that horny rascal, volunteered to take his place. "I know what you are planning," Dower said to Rooster. He continues, "Two things. One, sexually transmitted diseases are rampant here as are crabs and they are punishable by Article 15 for first offense. And two, the cost of loving has gone up considerably since the monopoly money fiasco. I hear it's five and twenty now [five dollars for a quickie and twenty for an all nighter].

"On another topic of concern, you guys need to know that there is a lot of tension right now between the RVN's and the Montagnards. Do not get between them or show any favorable inclinations to either side. The CO has requested a replacement for the RVN Captain, hoping to smooth things over. As you may have noticed the three interpreters are sticking close to the Ops office and the NCOs whenever we go anywhere.

"Wilson that damn monkey threw shit at me again, and he's wet and I think I will enjoy shooting him someday."

"Aw Sarge," Wilson replied, "He is only trying to tell you how much he cares."

First Sergeant Gross has suggested that I relocate to this hootch with you guys. He feels you need an NCO presence to monitor you. So after supper I am going to start moving.

Anyone want to give me a hand meet me at the NCO hootch after supper.

On our way to chow we saw May and some of the other girls helping Chief Poopie out of the compound. I was amazed that his body will tolerate this abuse every day.

We went through the serving line and Wilson saw this guy Cincinnati in the KP line. He was the guy who popped the grenade from his M-79 right in the middle of the patrol brief. After they acknowledged each other, Cincinnati reported he had been on KP and guard duty for the last three weeks. He said First Sgt. Gross is letting him join the patrol in the morning. He will be on point with an M-16. Not exactly a gift from the First Sgt. I know this fella will remember his mistake for a long time.

I decided to write home and the others got involved with their own projects. The night passed normally, two on, two off and then we greeted another day. It was early 11 December 1967. We began to think about the Christmas Holidays and missed home with our friends and family. This would be my first Christmas away. I felt sadness and loneliness, more so than since I'd been here. However, I have to move beyond that and prepare to face eight more months of separation. I have to get back home! I took a shower, ate some chow and practiced some Tai kwon do exercises.

The next morning Sgt. Dower informed me that I would be on the patrol this morning. The regular radioman was in sick bay and the back up was going on the re-supply run to Da Lat. He felt since I was Signal Corps trained person I should have no problem with a PRC-45 field radio.

"Look Sarge," I replied, "I did use one during our jungle training at Ft. Bragg. But that's been what, seven months ago?

Can they give me a few minutes with the back up guy before the convoy leaves? I could use a quick review."

I got forty minutes before we move out. So I grabbed a quick breakfast, stopped at Ops, checked out the radio, reviewed frequency and operational sequences and lugged it back to the hooch. Once there, I picked up my gear and reached down into my duffle bag for my personal pistol. I packed it away before shipping out last August. It's an S&W thirty eight with an 8 inch barrel. I had a leather holster and belt to go with it and three boxes of shells. I loaded the pistol, put cartridges into the belt and strapped it on. It is not as powerful as the Colt forty five but it may be an extra tool for use.

At the staging area I received four fragmentation and two white phosphorus grenades. So now I had a twenty three pound communications device, six grenades, an M-16 and flack jacket to carry the eight miles down and eight back from the Alpha Checkpoint.

"Listen up," First Sgt. Gross barked, "We are going to trek down to the check point this morning and on the way back set up an ambush site. I will lead one group and Sgt. Dower will be in charge of the second. We will remain in the field until around fifteen hundred hours, at which time we will regroup at the main highway near turn Four Alpha, which is here on your grid maps. Lieutenant Lee our ROK Liaison will be with us on this mission and he will travel with my group, as will the radioman, any questions?"

When no one responded, he continued, "The C-Rats are in those cartons in front of you. Grab however many you want and let's get going."

I looked up site and saw the convoy pulling out. Didn't they know we weren't down there yet? Maybe the Montagnards had already secured the roadway for them.

We loaded up and when we took the hard left just out of the compound I remember the trip we took to get to the downed Huey gun ship. I figured we would stay in the tree line until we broke into the tea plantation and follow the dirt trails down to the roadway. This was a meandering course on the downward incline. We maintained our separation. I was third in the formation: the point man, the First Sgt. and then radioman. I wondered if I'd remember the Zulu alphabet. To myself I recited: alpha, bravo, or is it baker, charlie, hell, I can't remember. I must brush up on that when we get back. In the meantime I hoped I wouldn't need to use it.

I felt clumsy and off balance. If I held my M-16 in the ready position, it slid to the microphone side. If I slung the M-16 to my left shoulder it seemed to sit better. About a quarter of the way down from the site we came upon a small opium field. With such attractive flowers; who would believe they could bring such misery. Drugs were plentiful and cheap! You could buy a garbage bag of hash for three dollars American. Some GIs had serious problems with it. Articles 15's were constantly handed out, even on our small compound. I had heard of guys being sent to stockade in Plieku for repeat offenses.

We neared the main roadway and at about mid point to our patrol zone between Alpha 4 and Pre Line Road. We stepped out of the vegetation and down to the roadway. First thing we saw were two ox carts carrying coffins, actually wooden body boxes, locally made. I mentioned to nobody in particular that the last convoy I went down on passed a few of them also. First Sergeant explained they lost a few locals down in the "Delta," ("I-Corps" area further south of us,) and were shipping them home.

The normal patrol formation spanned both sides of the roadway and staggered several feet apart to minimize injury to personnel in the event of a booby trap. The enemy loved using trip wires and unexploded shells or white phosphorous grenades and land mines too. We continued on to Alpha 4 and set a defense perimeter so we could take a break and eat. I opened a can of peanut butter and jelly sandwiches. I can't seem to get away from the peanut butter C-Rats. I washed down the four little two inch diameter delights with water from my canteen. I ate the candy bar and a can of pears. The cigarettes were Pall Mall this time. I guessed that the American Tobacco Company got the contract for their products when these meals were made. I tucked the smokes into my flak jacket pocket as we prepared to move out again.

We passed another ox cart coffin on the way to Da Lat. Further down the way we waved to some locals on scooters. Not much else came our way. We moved up into the brush and then the tree line. We saw pathways created by locals as well as VC. We set up our ambush there. We positioned the two groups in an angular pitch one left facing the other right which enabled coverage on both directions of the trail. We waited and waited some more. The ground cover helped us blend into the surroundings. I thought it a bit ironic that I carried a radio with a seven foot flexible antenna while trying to hide in three feet of overgrowth. I muted to minimize skip traffic or any other noises. This was a necessary inconvenience considering the radio would be our only line of communication in case of emergency.

First Sgt. Gross gave the word to move out. We returned to our patrol formation and traveled along the wood line until we came out on Pre Line Road. We turned up the incline and headed for the entrance to the compound. As we moved into the clearing near the gate, we were confronted by an unexpected situation. Twenty or so Montagnard troops argued with the RVN Captain. As we got closer we saw the Montagnards First Sergeant locked in the cage where they placed troops who needed correction. First Sgt. Gross dismissed us and told us to stay clear of the gate and the Montagnards camp. Clearly things were getting more intense. We took the First Sergeant's advice.

An Old West Remedy

The balance of the evening was a blend of angry exchanges and uneasy quiet. We were exhausted as we were up most of the night. I had one canteen full of coffee while on guard duty. As dawn broke the Montagnards seemed to be mollified. Their First Sergeant headed into the Operations Building. Soon after the CO and First Sgt. Gross were heading for the front gate. Gross yelled to me, "Get Dower and a detail to the gate double time!"

I thought this sounded like trouble. I ran to the Hooch to get Dower and some of the guys. Upon reaching the entrance gate we found the RVN Captain hanging from a tree limb near the cage. As the shock of the sight was settling in my mind, I noticed all three of our interpreters stood very close to us. They were Vietnamese as was the Captain.

We cut the rope and put the body in a body bag and placed it in a truck bed. By then the entire compound was awake and everyone, GI, Montagnard and Vietnamese workers were trying to figure out what happened. The CO posted a sentry at the truck to ensure the site was secure. With company coming and possibly some high ranking brass, we would need to be at peak awareness. We automatically concluded we would pull added security patrols and heightened alert at the site.

Sgt. Dower assigned Chin to guard the truck and body. Wilson, Rooster and I went along with Sergeant Dower. We manned the posts at our scheduled shifts. Later I heard the chopper, a Chinook, headed to our helipad. The dust kicked up from the prop wash of the Chinook and the pilots didn't drop the rear ramp until the engines stopped. An RVN Brigadier General and alongside is a U.S. Bird Colonel, a Major and two RVN Captains. I wondered if one of them was the replacement for the dead captain.

The CO and First Sgt. Gross saluted and greeted the visitors and escorted them into the Operations Building. A few minutes later one of the RVN officers motioned for the three cowering interpreters to come into the office. The door closed again.

After several days it was determined the Captain had committed suicide and no foul play was involved. Their First Sergeant was promoted to Lieutenant and placed in charge of the detachment. All seemed well now except for the interpreters who nearly wet their pants every time they were left alone.

Our favorite canine Shep disappeared. Rooster had us looking all over the site for him and no Shep! Rooster was disappointed and I found myself missing him too. I guess he was a good dog, fat little rascal that he was. The other member of our pet farm was still making friends. Freddie basically peed or pooped on anyone within range. Wilson was the only one who was safe.

Our Tai Kwon Do training went steadily. We were ready to take our first test for a yellow belt. Lieutenant Lee was a pretty cool instructor. We finally discovered why he ended up with our unit. He got to come along with us because of some trouble he got into in Nha Trang. It seems he got into an argument with a ROK MP Captain, must be a thing with MPs. Anyhow they were kicking and punching it out Tai Kwon Do style when Lieutenant Lee put the captain in hospital with broken bones and blinded him in one eye. He was given the option of returning to South Korea for a court martial or being assigned to a remote site away from the ROK's for the balance of his tour.

It was just over a week before Christmas, 1967. We were into this daily routine. There was no real excitement about the holiday season beyond the anticipation of a big feast. Rooster went back to Da Lat and got himself a new dog. He calls this one May. Whether that was intended to agitate our housekeeper or not, I'm uncertain. May was a moderate sized dog and she ate as much as Shep, which was as often as we put food in front of her.

We heard rumors of VC and NVA troop movement in our area. Some were concerned we would get hit on Christmas. You know, holiday morale crusher. The CO had the Mortar teams firing illumination rounds the past few nights just to let any potential encroachers know we were awake. Patrols were also increased along the berm area of the site. The CO wanted to see if we detected any signs of our defense perimeter being tampered with, concertina cut, or anything such as that. We found nothing that indicated any unwelcome visitors. This made me and, I think, everyone else a bit more confident in our security situation.

Lieutenant Lee agreed to get us to our green belts before the end of December. That meant twice a week training and practice sessions about every day. Hey, it wasn't like we had any pressing engagements anywhere at the time. We were stuck on top of a mountain in Southeast Asia and the training gave us a diversion. We set up a routine: sleep, patrol, eat, practice, eat and guard duty, with a little more sleep mixed in. The good part was the busier we were the faster time passed.

I got to go down on the convoy just before Christmas. Da Lat was abuzz with a lot of people, traffic and funerals. We saw two burials in the Catholic cemetery as we drove by. The RVN Academy was getting ready to graduate a class of new officers. I wondered if one of them was coming up to our site. I couldn't wait to get back and see if I got a package in the mail. I craved cookies or anything else sweet. I guessed I'd probably get my wish for Christmas.

The Montagnards were happy now. They invited Rooster over for supper one day. We wondered why the rest of us weren't invited. They served some stuff from a large ceramic pot. He said it stank up the place. He said it had fish heads, rice, what looked like cabbage, meat that tasted like chicken and spices that made it so potent that Chief Poopie kept busy burning our latrine barrels for a while. Luckily for him they also provided lots of strong rice wine.

One day as we were heading up Pre Line Road, returning from Da Lat. We passed a group of Montagnards heading out on patrol. We waved and they reciprocated, smiling. These guys are maybe five feet tall, stocky but small at the same time. I guess you could say they are about the same hue as our American Indian.

We entered the main gate and pulled up at the Operations Building. We dropped the mail and administrative stuff and headed over to the ammo dump and generator shed to unload those items. The last truck hauled the food stores and exchange items so they went up to the mess hall. That done we had a few minutes for ourselves before supper, I entered the hooch and saw May (the housekeeper) had finished my poncho rain coat. I tried it on. It fit well. Even the arm length was good. I wondered how she did that as she'd never measured me or anything. Maybe she patterned it off my fatigue shirts. Just about the time I made my conclusions on the rain coat, Chin walked in laughing. I asked, "What's funny Chin?"

He told me that Rooster was mad as hell at the Montagnards. He said he would have nothing more to do with them. Back when they invited him to dinner, it was their way of thanking him for their meal. "I don't understand what are you trying to tell me?"

He continued through his laughter, "Remember Shep? Well it appears he was part of the main dish, you know the other meat that tasted like chicken. They ate his dog and wanted to share their good fortune."

"Wow, he really loved that dog; I can understand his being upset." "Yeah, and now he's got to keep a close watch on the new dog May or he may get another dinner invitation. I know I just couldn't help it. I will try to control myself when he's near.

During mail call I got one package and six letters. I carried my loot back to the hooch. The temptation to read them was overwhelming. I was always a sap for anything wrapped or sealed. I decided I would know soon enough. The excitement could last a bit longer if I wait until after supper. There, they'll be fine in my foot locker. I headed for the mess hall and joined my hooch mates in the serving line. We found a table, sat down and started eating.

Sgt. Dower joined us for chow. He cleared his throat before making his announcement, "You guy's have all just been promoted to PFC (Private First Class), by the order of LBJ and this man's U.S. Army. Congratulations!"

Thank you Sgt, I replied, may I run home and tell my wife about this? Then the others clowns chimed in, Yeah, us too!

"Not a chance of that happening. You are now even more essential to the mission." Everyone got a good laugh at that.

We finished our meal and headed back to our little piece of Pre Line Mountain.

I opened my foot locker, pulled out the letters and placed them in order of postmark dates; most were posted during early December 1967, not bad for the distance they had to travel. They were all from Sherrie. I savored the words as I read through them, trying to smell my wife's perfume through the paper. I imagined her voice and her face as I read her words. I missed her more than I could articulate. After the letters, I opened the package. Inside a tin container were the remnants of a few dozen chocolate chip cookies. They still tasted good.

It's Christmas day, no presents to hand out, so we just spent the day as usual. The meal though, was fantastic. There was every dish that you would expect to see, turkey, and ham, mashed potato's, sweet yams, green beans, rolls and cranberry sauce. And more types of dessert than you can name.

Just after Christmas, we tested for our next belt. I advanced to green belt and got another package from home. This one had a cake in it and a can of spinach. My mother-in-law knows I hate spinach, I ate every bite! When you are awake most of the night you get hungry. Spinach isn't that bad, especially at two in the morning.

A few of the guys would be leaving soon. Their tour was over and they would welcome in the New Year stateside. I was only halfway through my stay.

Chin, Wilson, Rooster and I were promoted to E-4. It amazed me that I stayed an E-2 for over a year and in the past few weeks I received two promotions. I was thankful for any pay bump so my wife could get a larger monthly allotment. We might be above the food stamp level now.

Prelude to Tet

C hin went down to Da Lat on a resupply run. He said there was a lot of activity in town due to the upcoming Tet holiday on January 31. Everyone in town was ready for a celebration. I had no idea what it's about, some Lunar New Year thing, I think. I couldn't tell yet if the mountain folks celebrated it. Chin said that two of our three Vietnamese interpreters went down on the run and did not return with them. He figures they are on holiday leave. They retrieved a lot of mail on this particular run. The Christmas backlog has caught up.

I got several cards, two packages and a bunch of letters to read. Package number one is a cake, smells like a spice cake. It's wrapped in tin foil and inside another box for protection. Somebody must have really tried to get it here in one piece. They probably didn't realize they had to contend with two postal entities though. At least I didn't need a knife. I offered to share and multiple hands reached in for a scoop. It tasted excellent and everybody, including me, was surprised to see that it was made with tomatoes. A tomato cake, from my best friend Ron's mother. She enclosed a note which in addition to saying the cake contained tomatoes, read, "Merry Christmas and be careful."

Things had been relatively quiet lately. Patrols seemed more relaxed and I played radioman for several of them. The Montagnards were just a bit antsy and uncomfortable with the lack of activity. I suspected they anticipated a major ambush if they let down their level of awareness. These little fellows were fierce and determined warriors. I had only heard positive comments from the other troops about them. Gladly, they were on our side of this crazy war, conflict, or intervention whatever.

I was ready to go home and be with my family and watch my baby being born. Oh, I could be on R&R right now, in Australia or Japan or Hawaii. Problem is she couldn't travel so far in her seventh month. So, I stayed in country and hoped they might take the thirty days off at the end of my tour.

Sgt. Dower told us that the Montagnards wanted to set up a couple of ambush sites tonight. We would support them with illumination flares. Local information sources provided intelligence on VC movement in the area near Da Lat. We spent the rest of the day and most of the evening doing menial tasks and some personal stuff. There was nothing essential going on and things on the compound were as normal as they can be. We ate supper and cat napped until guard duty rotations pulled us back to war.

The night was dark, maybe due to the new moon and the Vietnamese lunar thing. I was at my post and the mortar pit guys stared firing illumination flares in the sky over us. Wait! What's this? I saw red streaks coming in the direction of the site. Whoa!! Where was that coming from? Bam, boom, bam boom and again, bam, boom. Three rockets smacked the berm around Lima 2 bunker. Chin was in there.

Before I could finish the thought I saw three more red trails of incoming rocket fire. This time they fall inside the compound itself. Wilson slid into the bunker with me panicked, "What the shit is going on?" I'm not sure yet. I do know that Chin's in Lima 2 and the first rockets hit the berm near him. Yeah, Wilson said, Dower and Rooster are on the way there now.

Lieutenant Lee jumped into our bunker. I noticed he didn't have a flak jacket or helmet, just his weapon and ammo belts. He told us Chin was fine. The impact was right of him and about middle of the berm. The perimeter took the hit though and they may be able to penetrate our defenses if they have any sappers with them. This is a common tactic. They carry satchels of explosives into the defense perimeter and blow them up to neutralize the claymores and bouncing Betty's. It usually blows out sections of concertina too. If that happened, they might gain access to the compound.

While we pondered this we heard automatic and semi automatic small arms fire and saw the tracers (every seventh round is a tracer) flying back and forth in the wood line below. Our Montagnards have engaged the bad guys. We can't fire into the area for fear of shooting one of our own. So we sit there and watch the fire fight, anxious to get involved, and at the same time relieved that we could not. More flares went up and the sky was alight with the eerie red/white glow. The fire fight continued for about an hour and then slowly ebbed to sporadic shooting and then to none. All went quiet but the flares still lit up the countryside. We waited out the darkness of night and anticipated the dawn and what might be revealed.

Daylight arrived and we surveyed the terrain where the fighting took place. The berm was a bit disturbed and one of the claymores blew, probably struck by a rocket. There was no indication that anything happened below us at the wood line. It would be a while before we found out any information on what the Montagnards found, shot at and/or killed.

Dower said all but one man could stand down. The rest could go to chow down and then someone could relieve the guy who stayed, in order for him to go eat. I volunteered to stay on and the others left the bunker. I gazed back toward Lima 2 and wondered. What if there had been sappers? What if they were planning on overtaking us? Could we have held them off? We had no aerial support last night. Had anybody requested it? I lit up and took a few deep breaths. Man it was good to be alive.

The days immediately before the Tet Holiday were trying. The powers that be were negotiating a temporary truce for the holiday. Every night we got rounds incoming. We fired back and put up flares. We got some sniper fire, but no hits. There was no major damage, no injuries and no sleep. The mortar pit guys were getting a work out.

By day we kept low and took minimal trips around the compound, especially not out in the open areas. The villagers stopped showing up for work. As darkness descended, we took mortar rounds sporadically. Most fell short of the essential areas and those which did enter the compound did minimal damage to equipment or personnel. We sent up flares to detect troop movement. We knew somebody was out there, we just couldn't see them.

One night, 26 January1968, at around 0300, the CO told the mortar crew to fire off a couple HE rounds. They complied almost immediately. The only problem was they forgot to reset the tube. The round had already dropped into the tube when the pit boss realized the mistake. This round left the tube in vertical trajectory (straight up). It was reasonable to conclude it will come down in the same fashion. They cleared the pit and warned the commo shack of an impending explosion. The pit got a direct hit and the commo tower developed a sudden and precarious tilt that shut our line of sight (VHF) and microwave (UHF) systems down abruptly. We still had

PRC-45 capacity and we tapped into the Page Transmitters within a few hours. It wasn't until later that those of us who were in the bunkers found out our own guys did it.

Now things got even more nerve wracking, no mortar pit meant no illumination flares. With no flare capability, we had to wait for the bouncing Betties to get tripped and concentrate fire on that specific area if infiltrated. We were fortunate in two separate areas this night. First, this was only a harassment plan for the VC and there was no plan for an assault. Second, and more significantly, they had no idea we had compromised our communications and mortar defense systems in one blundering move.

We sat out the remaining darkness straining to see and once more hoping for dawn to break quickly. Finally, sunlight cracked the darkness of night and we saw what was out there. We stayed posted and Sgt. Dower brought around C-Rations. He apologized, "Coffee will be a little late today guys you have to wash it down with water or make some hot chocolate from your c-rats cocoa mix." He related that the pit will be operational by noon and the tower repaired by the next afternoon.

The local labor force returned from their unannounced time off. Our interpreter questioned Chief Poopie about their absence yesterday. The VC ordered the villagers not to go to the compound. They naturally knew the punishment for disobedience would be dead leadership. Since Chief Poopie was the village chief, the decision was not difficult for him.

I ate lunch and grabbed some coffee, added three sugars and heavy cream. I needed that caffeine boost and lit a Lucky Strike I got with the C-Rats at breakfast.

I took a nap and a quick shower before supper. I got up from my nap and our housekeeper, May sat, actually squatted, near the doorway with everyone's boots, polishing away. Rooster's new dog May lay near the shower, in the sun. She lifted up her head as I approached, wagged her tail and lay back down. Freddie was not so welcoming. He squealed, grabbed a pile of poop and flung it. "Not even close you scum bag," I said. He didn't understand or care, but it gave me some satisfaction to yell at him.

After supper, I dropped off to sleep again and was awakened by the sound of metal clanging on the floor. Lieutenant Lee was mounting a sniper scope on his weapon and had dropped a screwdriver. He apologized for waking me. I asked about the scope.

With a sly grin Lieutenant Lee responded, "I think I will go out tomorrow and see if I can find any good targets wandering through the area. Maybe I'll get lucky and find an opportunity to test my scope.

Trip to Da Lat

A re-supply run was set so we could pick up another mortar tube. The CO wanted to build a new pit away from the commo shack and use the original for flares only. I understood his reasoning after the previous night's experience.

I would be radioman for the run. We moved out at first light with no layover. Intelligence had it that a heavy movement of troops was coming in from the North. The idea was to build up troop strength during the Tet truce period, which would begin in two days. A large number of the RVN Army was already headed for their hometowns. They expected more than half the armed forces of South Vietnam to be on holiday by 31 January.

A few hours before we left, I realized I didn't have much time to get my stuff together. I concentrated on that issue after I ate. Walking back to the hooch Chin excitedly informed me he would drive the follow jeep. I was in the lead jeep with the First Sgt and his driver, a guy named Walters from Atlanta. He planned to return to school when he got out. He wanted to be an accountant. The GI Bill would pay for most, if not all of his college and he has only six weeks left here and four months left to serve when he gets home. They will probably discharge him when he gets to the States.

I completed the radio test and climbed into the back seat of the jeep. First Sgt. Gross climbed into the front and we took off. We traveled slowly toward the front entrance and then swung out onto the trail leading to the main highway. We went about fifteen miles per hour on this pothole riddled narrow dirt surface. We reached the highway, turned left and accelerated to about thirty. I saw Chin as he made the turn and we are all in queue.

Traffic into Da Lat was heavy. The locals were slow to yield to our horns and rapid speed. We had to dodge but had no collisions. We swung around and saw a flat bed truck hauling three caskets. I guess these folks wouldn't be celebrating Tet this year or ever again.

Da Lat was a gorgeous city with trees and flowering plants which acted as decorations for the upcoming holiday. It was spectacular. Walters turned into the compound and slid to a stop in front of the HQ office. The trucks headed for the supply depot. Chin pulled up behind our jeep. His black hair stuck out from under his jungle hat. It was now a dusty tan from the trip. I sarcastically asked him if he is trying to camouflage himself to look Asian. He told me, "Kiss his Asian ass." He called first dibs on the shower when we got back.

We loaded up within an hour and started the trip back. There was tension all around, nothing I could nail down, just that feeling of something impending. People stared at us as we passed by. They paused and their eyes followed us until we were well past them. Pedestrian traffic was heavy due to the holiday. The right turn to Pre Line Road started us up to the compound and we were back in time for lunch.

We unloaded the perishables before we ate. We would offload the rest after chow. I snatched the mailbag and dropped it in the mail room. I returned to the hooch and stowed the PRC-45 after making sure I turned it off. Then I headed up to the mess hall.

After lunch, other troops gave us a hand unloading the fuel barrels and ammunition. We lugged the boxes of mortar rounds up to the repaired pit and set the tube in place. Sgt. Dower instructed us to stow some of the rounds down near the new pit which would be ready by nightfall. The new pit was located about dead center between the inner gate and the Operations Building. How did we not see it when we came back from the resupply run? There are three deep holes spaced at six, nine and twelve o'clock within the pit. It was heavily sandbagged with a three by three foot steel cover for the opening where the rounds will be stored. Sandbags would also be put on top of the steel plates for protection. Walters lugged the new mortar tube and stand over and began to set it up. Before I knew it the pit was ready for use and none too soon.

The CO called a staff meeting that night 29 January 1968. Apparently new intelligence identified some concentrations of NVA troops in II and III Corps areas of the country. They would probably do some speculative planning as to what was going on and how it is to affect this area of II Corp. We waited until Sgt. Dower got back from the briefing to hear the details. I decided to practice some of my Tai Kwon Do patterns. Lieutenant Lee was after us to discipline ourselves and set up a regular training regime.

On the way back to the hooch I passed Rooster. He and May (the dog) were going for a walk. He had her on a rope. She was typically secured to her little house that sat between the hooch and the place where Chief Poopie took his afternoon nap. Rooster was still upset about having Shep for dinner so he took no chances with May. Maybe he didn't realize that as soon as he returned to the States, she would be in the pot.

On 31 January 1968, the Tet Holiday, a truce was announced. The CO warned us to remain sharp. They could be setting up something to break during or after the holiday. None of our local workers would come on the compound for the next few days. We took over their jobs. That meant KP, latrine duty and laundry in addition to guard duty and routine patrols. Wilson responded to the increased work load with, "Jeeze, you'd think we are in a war or something, this is going to be hard on me."

Suppertime rolled around and afterwards we cleaned our weapons and secured the gear we would use on post and made sure we had a few packs of c-rats in the backpack.

I was restless that night. Maybe I ate too much supper and then pigged out on some goodies from home. I went over to relieve Chin at close to 0200 hrs. As I walked to the bunker I heard noises that sounded like thunder. Off in the distant sky I saw brief flashes. I listened a little closer. I arrived at the bunker and Chin looked up to see me and said, "Hey "Simp" what gives, you hear that?"

"Yeah, I thought it might be thunder. Now I think it may be mortar or light artillery over toward Da Lat. Let me go back to the hooch and get my PRC-45, I can tune in and see what's happening."

Excitement gripped me at the same time as a feeling of uncertainty. Could Da Lat be under attack? No, it was probably an engagement north of there, possibly an ambush. No, I don't think there would be any light artillery on an ambush. I snatched the radio up and stumbled on Rooster's bunk. "Sorry man, I tripped on your bunk."

I retuned to the bunker, slid onto the bench and turned the volume up. I fine tuned the operational frequencies we used for the II Corps area. We heard the alert loud and clear, "Mayday, Mayday! Da Lat Command under heavy artillery and mortar attack. Sizeable force of Victor Charlie troops are moving into the city. Two MP Posts overrun, some casualties, request air support and ground reinforcements."

Chin and I looked at one another in the darkness of our bunker. I was overwhelmed with apprehension or fear or maybe plain disbelief. Whatever it was it jolted me to full awareness. I ran for the Operations Building and just as I arrived there, First Sgt. Gross opened the door. I updated him on the radio message. He yelled, "Get your guys up and posted, tell Dower to take your sector and keep your radio on!"

I headed back for the hooch to wake the guys. They were already up and heading to their positions. I then headed back to Chin in the bunker. The whole compound came to life. The Montagnards were ready to deploy a couple of squads off the compound. We were in full lock down which meant no lights, no talking, no anything except eyeballs straining to see through the darkness. The pit crews started firing illumination flares, but before the first one popped we got incoming small arms fire. Once the flares ignited, I saw them. The Viet Cong are climbing up to the perimeter of the berm.

The Tet Offensive

The new pit opened up with white phosphorous, while Pit 1 stayed with the flares. The first round hit and I saw the brilliant glow of the phosphorous spray in many directions, beautiful and deadly. As it exploded I heard the screams of pain, agony, certain death to those directly in the strike zone. I didn't have time to consider their misfortune though. My mind was on survival, mine and my fellow soldiers.

We fired short and deliberate bursts. I aimed the M-60 right on a group and pulled the trigger. Several fell, more surged forward. I fired again. We all did. Same scenario—some went down, others took their places. Illumination flares reveal hundreds maybe thousands of them. My radio crackled with a request for air support. I heard the coordinates for our location and shout to the guys that we are going to get air support soon.

Again my radio came to life. Bravo Sector, Bravo 2 bunker advised they had visual on intruders advancing up their berm area and requested mortar support. The CO directed Pit 1 to alternate flares and HE rounds and assigned the new Pit to support the Bravo sector in similar fashion. Suddenly there are the sounds of impact and explosions on our site. Small rockets (more nuisance than anything) and some mortar rounds that could inflict damage were hitting us. The bunkers were the safest place to be. As long as the ammo held we wouldn't risk going out.

Sappers reached the first perimeter and we concentrated fire power on them. They carried satchels of explosives and attempted to blow a pathway for their troops. These folks are the VC version of the Japanese suicide pilots. They were usually high on heroin or some other drugs. We had heard stories of them tying off their limbs in case they are blown off so they could continue on their stubs to get the job done. Supposedly they wouldn't bleed out as fast and it allowed them to get closer to the target. My M-60 machine gun took a bead on two Sappers. One went down; I saw the other's head explode. I concentrated on the one who is down.

Wilson dropped a grenade projectile in his M-79 and fired it off. The WP (white phosphorous) hit about ten yards from the target. The flash told us all we need to know, the screams evidenced the final moments of his torment.

Forty five minutes passed and Da Lat was still taking a real beating. They lost control of major areas of the city and had considerable casualties. The main compound was secure and several off site areas were still being defended. I kept hearing the call for air support and the reply was always the same, "Deployable aircraft delayed, naval air support being alerted." Cripes, what the hell is going on? How can the Air Force and Army aircraft be too busy to help them, and more importantly, US! If they had no aircraft, how will reinforcement troops get to them?

Another round of incoming brought my thoughts back to real time. Another surge of troops started up toward us. Not much for us to do other than continue with our prior efforts, lock and load, fire at will and hit what we were shooting at. Where did all these people come from and how did they get here without somebody seeing them? Sappers on the berm! Ka boom! He set it off and we couldn't see the damage that may have resulted. I threw the M-60 over toward the area and opened up. Another flare lights up the night. The sappers work failed to destroy the wire barrier. We couldn't let them get this close again. Chin yelled something and started rapid fire with his M-16 on the left side. It sounded like the fourth of July. More lead went in that guy than went into the water buffalo in Da Nang. He had no chance to get his satchel to blow.

Sgt. Dower returned to see how we are holding up. I told him the PRC-45 transmissions didn't sound too encouraging. I also told him we'd need some more ammo. Chin volunteered to replenish our supply from the ammo bunker at the rear of our compound, a good distance from us.

We realized that we may be stuck in the bunker for a while and that we also needed to conserve our ammunition. No more automatic fire on the M-16's and short bursts from the M-60 from here on, unless we were being overrun. Word passed to all the bunker positions by Dower and the other NCOs. Lieutenant Lee joined us in the bunker and told us that he was at the berm doing some sniper work. How's the hunting Lieutenant, I asked? He was sure he popped a couple of them, guess we will have to wait until morning for confirmed kills.

At close to 0430 hours with daylight soon arriving, the firing ebbed to just small arms fire. This either meant they were regrouping for another assault or that they are sneaking back into their hiding places and awaiting darkness again. I'm personally hoping for the latter.

Finally, the sun rose. We saw the area clearly. Other than the two sapper's remains, there were no bodies to be found.

Perhaps they removed any KIAs in order for us not to get a count. Generally, you could estimate the force strength by a body count, it's arbitrary but it can be fairly close. Sgt. Dower came over with Chin and Lieutenant Lee, hauling, I should say dragging, several boxes of ammo. We grabbed something to eat and then patrolled along the berm. We needed to check for any breaks or booby traps and replace some of the concertina wire.

I was still monitoring the activity in Da Lat. They were still under heavy assault. We couldn't help them yet. We needed to establish or own circumstances and the potential of even getting off this compound. It would do us no good to get ambushed on the road down and reduce our own manpower levels. Brown was down there and we had been through a lot of this stuff together. Wilson jokingly comments that Brown is probably beating them to death with his pots and pans. We chuckled, ate, cleaned out the bunkers, restocked them with ammo and rations and then waited.

The Montagnards had been out for about two hours now and it was around 0930 hours. We hadn't heard any gunfire or anything. The CO ordered us to take two squads out and check the berm and perimeter. For once the position of radioman paid off. I don't have to carry any concertina spools, trip wire or bouncing Betties. They did give me four claymores stuffed into a sling bandolier. The idea is that one squad posted while the other worked. Rooster, Chin and Wilson are the unfortunate workers, along with several others from the outback hooch's.

While on our rounds we encountered blood soaked and flattened ground and bits of clothing and flesh. It was evidence of dead or wounded that have been retrieved in the darkness. The only bodies are the remains of those that got into the berm and concertina. I was glad I didn't have to remove them, trip wires were hard to see in the daylight and impossible at night. We checked our drawings of the layout to ensure we knew the locations of the deadly little boogers. It took a little more than two hours to go around and complete the repairs.

Upon returning to the compound we saw a convoy of trucks sitting in line ready to roll out. First Sgt. Gross checked with the CO and laid out the plan for us. We were going to start down the mountain road with a combined group of GI's and Montagnards to see how close we could get to the city and also try to gauge the size of the forces still in the area. The most recent intelligence we had was that access was blocked by highway and they had control of the airstrip. That pretty well eliminated any resupply except by air and it was hot for helicopter traffic due to small arms fire. The troops were holding their own and under heavy assault, the main compound and the Academy sustained damage but were still holding as were a couple of remote sites staffed with MP's. The question was how long could they sustain before being overrun?

I put fresh batteries in the PRC-45. It had been on for a while as we are trying to keep aware of the situation in Da Lat. I also overheard some traffic about Hue and Saigon being stormed. I wondered if they planned on honoring the truce at all or if this was only a ploy to get everyone to let down their guard?

Our security patrol trucks left in convoy. We proceeded with caution, wary of roadside booby traps and potential ambush sites, not only on the trail down the mountain, but also the highway into Da Lat. Once we got to the roadway we dismounted and deployed into our patrol formation of staggered positioning along both sides of the roadway. The deuce and a half's followed about three hundred feet behind. The Montagnards deployed into the wooded embankments and hills on both sides of the roadway.

We traveled about a quarter mile and saw nothing. After another quarter mile we reached one of the RVN Checkpoints. It was destroyed. We also found the bodies of nine RVN troops. We decided not to take the bodies with us. Instead we gathered them up, placed them in body bags and lay them out at the edge of the wood line. During that hour our Montagnard partners advanced ahead of us. We resumed our trek toward the next security post and saw the Montagnards just as we got hit. The deep rumble of the weapon confirmed what someone yelled, "Incoming! Heavy!"

It seemed to come from a thirty to fifty caliber weapon. The rounds will cut three and four inch trees like they were blades of grass. We dove, scattered and yelled in unison. I crawled into a small depression on the side of the roadway. I cursed the PRC-45 I carried because I couldn't get as flat as I wanted to nor could I get a comfortable firing position laying flat on my belly.

We had to find out where the firing was coming from so we could at least return fire. They were on a small knoll, densely covered in small trees and brush, at about the five o'clock position. Rooster loaded a grenade into his M-79 and fired, loaded and fired again. It struck short of the target area. He loaded a WP round, elevated his weapon a tad and fired. It hit closer and the spray of white phosphorous penetrated the area near the target. Everyone fired toward the enemy position.

When First Sgt. Gross called for the radio I crawled to his side. He radioed in what happened and I stayed near him. He was a big guy and more of me was protected by him! Soon our attention was diverted to the trucks, which were between us and the guys shooting at us. Suddenly the firefight was over. They left as quickly as they showed up. Their ability to meld back into the surroundings was uncanny, especially with weapons and casualties.

We carefully surveyed the area and assessed damages. No wounded or killed some scratches and bruises for sure, they will heal. We were still at least four miles from Da Lat and decided to return to the site. First Sgt. Gross advised the CO of conditions and that information was passed on to II Corps for intelligence updates.

For the next two days and nights we got assaulted and had more firefights. With each trip we got closer and the resistance was more intense. The second day we lost one and three were wounded. They were Medivaced out from the compound because we could not set up an LZ on the roadway. We were running low on mortar rounds and small arms ammo. The CO requested resupply and a Chinook brought in four large skids of assorted goodies.

The third day we made it to Check Point Bravo, about a mile and a half from Da Lat. They (the bad guys) had destroyed it totally. Eleven dead bodies were already in decomposition from being in the open for several days. The odor was stifling. Another benefit of being radioman was that the other guys had to bag the remains. The victims were mutilated, bayoneted, eviscerated or shot multiple times or all of the above. The firefight that day was pretty feeble compared to the day before. Perhaps they were suffering high casualties in Da Lat or concentrating their major efforts there. Our major opposition was sniper fire and occasional heavy automatic fire.

By day five, 4 February 1968, we were within a mile or so of Da Lat and reinforcements took our mission over. We went back to the site. These past few days had taken a toll on everyone. We were dirty, tired and emotionally weary. A respite would be welcome. Unfortunately, that is not going to happen tonight.

We Don't Rest

As soon as dusk arrived, so did the heaviest assault we experienced. They hit us from three sides simultaneously. Both mortar pits were in rapid fire posture with alternating HE, WP and Illumination rounds. They were inside the first perimeter on the Lima Sector. We threw as much defensive fire as we could and they just kept coming. From about 1930 to 2300 hours we kept them back. Chin and I had seen several Betty's deployed and we were holding the claymore wires and blowing them along the first ring randomly. The CO requested air support for our perimeter but we had no deployable aircraft in our area of operation at the time. At approximately 0130 hours rockets were hitting all over the compound. The CO put out a second request and got aerial support assigned to a C-47 gun ship (referred to as Puff the Magic Dragon). The pilot advised he was about ten minutes from us and would do a pass around the site.

We held our own. The Montagnards reinforced our side of the perimeter. They fought along the open berm with no protection as we had in the bunkers. They seemed fearless as they rose up to fire their weapons and dropped down when done. A thud and then explosion shook our connex bunker. It must have been really close. Sand and rust shook and fell on us inside. I wondered if our little friends escaped injury. I heard yelling, then the unmistakable cry of the wounded;

you almost feel the agony and pain. I picked up the PRC-45 handset and yelled, "MEDIC, Lima 4, Medic!"

Chin moved to the opening and I heard him say, Wilson its Wilson, God! He's bleeding man, Simp! Do something."

"Chin, get back to your post. We can't stop now, they're too close. Medic is coming. Doc will take care of him." I was amazed at my own calm assessment of the situation. But then I turned and fired my M-60 at shadows. My rage overtook my common sense as I screamed, "Wilson! You hurt Wilson. Take this you bastards!"

Rooster stuck his head into the bunker, "Simp, Wilson's hurt bad man. You got to help him! Please, Simp, I think he's gonna die!" Chin looked at me and moved to the 60 so I could help Wilson. I nodded, grabbed the first aid kit and left the doorway of the bunker. Wilson was on the ground about three feet from his bunker. Rooster had him cradled in his lap. He said, "Hang on man, hang on, Simp's coming."

I saw his upper torso in the flare light. His face and neck were bleeding so much the blood had saturated his flack jacket and fatigue sleeve on the left side. I got closer and saw that part of his face was missing. His jaw bone was exposed and some teeth were missing and he had numerous shrapnel wounds to his left torso. I compressed the opening in his face in an attempt to stop or slow down the blood loss, at least until Doc arrived with an IV. I took out a third dressing and pressed it down hard. Wilson screamed in pain. I tilted his head toward me so the blood would not choke him. I told him to slow down his breathing and try to calm down.

I held him and turned to Rooster, "Rooster, you need to get back to you position. I'll stay with him until Doc gets here."

Rooster protested, "I can't leave Simp, he's my partner."

"Yeah, I know that, but if the VC gets through, nothing we are doing is going to save him." He stared at me, calculating my words and then gently placed Wilson into my lap and returned to his bunker. Wilson moaned and grasped my arm. I whispered, "Its okay man. I won't let you die." Just as I said those words, his grip loosened and he seemed to become less anxious and tense.

Sgt. Dower and Lieutenant Lee arrived together. Lieutenant Lee went into the bunker to help Rooster. Dower looked at me for my assessment of Wilson's injuries. I just said he will survive. Seconds later Doc arrived and started an IV. He loaded Wilson onto a stretcher and he and Dower took him to the aid shack. I turned to go back to the bunker and saw a flare drop from the sky. I wondered where it came from.

Next I saw the flashing of red tracers and the sky erupted in a deafening roar. It was Puff firing on the enemy troops it looked as if there was a steady rain of tracers when in actuality every seventh round was a tracer. These guns put out rounds so fast; the illusion was there are not six rounds in between. Victor Charlie and his friends were finding that out the hard way. Puff made the first pass around our site and we just marveled at the awesome lethal weapon circling above us. I was glad these guys were on my side of the battle. Occasionally I saw a green tracer shooting up at the plane. Those dummies didn't know they were giving their positions away to the air crew, who in turn drop thousands of rounds in their direction. After less than a half hour of Puff, the hostile fire ceased. The bad guys packed up and apparently went home.

We waited out the rest of the night for word on Wilson's condition. No word.

Long anticipated sunrise was welcomed by us all. The landscape was a sight. Our shower was severely damaged with numerous holes in the ground from impacted rounds, and several body bags sat at the aid shack. Outside the berm the scene was dramatically different. There must have been more than a hundred bodies, some inside the second perimeter. We will have to get the repairs done quickly, hoping the snipers left with the rest of them scumbags. Radio traffic still advised Da Lat under siege, troops holding out and reinforcements were deploying into the area. Resistance was still strong in some sectors of the city and they were going street by street, house to house. That's the nastiest type of fighting, booby traps and ambushes are so easy to set up.

Hunger pangs could disrupt even platonic thoughts of peace and sleep and other good things. So, C-rations filled the empty and growling monster inside. Chin and I cleaned out the bunker, identified which claymores we set off and detached the wires from the bundle, number three, four and five. I stepped out of the bunker to take a leak and light a cigarette. Rooster and Lieutenant Lee were just completing their clean up and also stepped into the morning sun. It was warm and we all stretched our tired muscles and attempted to limber up.

Lieutenant Lee nods a greeting, I nod back. He headed toward the hooch and Rooster walked to my bunker and said, "Simp, Man I was scared last night, I was even praying. I asked God to save me and to save Wilson and everybody I could think of, even the Montagnards who ate my dog." You were not the only one scared man, we all had it. Maybe fear is good. Maybe it makes you more determined to fight hard, do stupid things, and take chances. I think I heard someone once say, "A hero and a horse's ass are only separated by a split second in battle."

You did a brave thing for Wilson last night Rooster, you probably saved his life.

"Naw, you saved his life Simp, I just had to convince you to do it. I said "Let's go see if we can find out how he's doing. I'm pretty sure Doc can use a break too. We may be able to give him a hand." Chin headed over with us.

We walked only a few steps before we saw Freddie the poop throwing, pee spraying, screeching pest of a monkey. His little body lay still. His eyes were open, but he could no longer see as he was another casualty of the war. I got back to the bunker and grabbed an empty claymore box. Chin gently lifted Freddie's little body and placed it in the box. We decided not to tell Wilson until he gets better. We buried Freddie and then went to see Wilson at the aid shack. Doc was about thirty deep in guys sitting inside the makeshift triage area. He stabilized some, changed IV's on others. We came in and asked about Wilson.

Doc told us, "Wilson's in the shack, sleeping. I gave him something to dull the pain and he's resting okay. He won't win any beauty contests from now on. He will survive though." He paused as we reacted to the great news. He continued, "See if you can give these guys that can walk and move around some water and maybe find some C-rats for them. Simp, I can use you to help to monitor the IV's and I will take on the critical. Roger that Doc, I just want to look in on Wilson first.

We stuck our collective heads into the aid shack to find Wilson propped between two tables. His head looked two sizes larger than normal. Bandages covered his face and eyes. From the neck up he resembled a mummy, straight out of Egypt. At least he won't ask us about his monkey, I thought. I asked Doc about the KIAs, seven GI's, one was supposed to leave soon, going home, Walters, his name is Walters, Doc says. Still going home, not the way he intended none of them expected this. I don't want to think about it, you feel sorrow, anger and relief at the same time. A friend is lost, "why" is the question you ask? And the relief is the selfish thought, "it wasn't me," I'm still alive. I moved along the line, checking IV's and helping some to their feet, they needed to walk, move, revive their bodies; sure it must hurt, still got to get them up.

I looked up at an empty sky and heard the Chinook before I saw it. When it came into vision, a big red cross covered the whole thing. Medivac was on the way in. These guys would head to Cam Rahn Bay for medical treatment, then Germany or back stateside, depending on how severe their injuries were. Doc yelled to me to round up the walking wounded and head for the helipad. Those who couldn't walk went on litters. We got them on board. Wilson was still unconscious. We said our goodbyes and watched it leave and wondered if replacements were coming in today. Everybody needed replacements, a number of units have lost troops throughout the country. We were probably well down on the priority list for filling vacancies.

The next several days and nights went by with varying periods of skirmishes, nighttime mortar attacks and random firefights. Apparently the Puff attack broke their will or decimated their strength. I did hear on the PRC-45, (now my constant companion), that Da Lat was retaken and the VC Battalion retreated. There were still small pockets of resistance from local VC groups, but nothing like the Tet assault.

Time to Chill

The local work force showed up today, another indication that the enemy presence was greatly reduced, and things are returning to some semblance of pre- 31 January activities. This was good. The C-Rations were getting really old three times a day and it was getting difficult to find clothes that didn't reek. We took showers wearing our fatigues. We soaped them up and then washed ourselves.

Supplies came by air and we got a couple Chinooks in weekly. No mail for a while, the main postal spot was in Saigon and it was one of the major cities hit during Tet. The old capitol city of Hue was the only place that we have not recaptured. Lt Derf and Cherico were there, and some others from the old 337th Signal Company we shipped over with. I hoped the best for them. I heard on AFRVN (Armed Forces Radio Vietnam), that they (the Viet Cong) coordinated the attack on five major cities on January 31st. They also said casualties were high for the enemy forces, better than ten to one. If you take away the inflated BS of the propaganda, it's probably closer to five to one, my guess. Time and body counts will tell.

We got the shower tower back up at our hooch and back in service. At lunch time we anxiously tasted the warm, mostly real food again. We reach the mess hall and chow down on fried chicken, mashed potatoes, green peas, peaches and applesauce. Not bad, considering what we'd been eating for the last few weeks. I could only imagine when Saigon got hit. The Air Force guys had to give up A/C, cold drinks, real beds and hot food for several days. Life must have been really bad, yuk-yuk.

Wilson went back to the states via a Navy Hospital ship for the reconstructive surgery he needs. His war was over. He would probably be discharged when he got out of the hospital. When we got his mailing address at Walter Reed Hospital in DC we sent letters, none of us ever heard back from him. It hurts to think he didn't write back, he was our "Brother".

It was the third week of February now. First Sgt. Gross announced we'd be taking a convoy to Da Lat in the morning. The majority will be going to assist in the rebuild and recovery effort at the compound in Da Lat and the four check points we passed by on convoy. I started thinking, which was not the right thing to do in this man's Army. Here we were down by thirty men. That left sixty two, minus Officers and NCOs which left fifty one grunts. That means we have a better than two to one chance of going to Da Lat. I looked around and figured we were all going to Da Lat.

That night was really weird, no war noises, explosions, weapons fire, just eerie quiet. It had been some time since it was so peaceful. It made me a bit apprehensive. Nonetheless, nighttime faded into dawn and we were relieved from our posts. I headed for a shower, then food. After breakfast I headed to the Operations Office for a briefing from the CO on what, when, where and why. We were taking thirty down; there goes my calculations, its now three to one odds. I headed back for the radio and my sidearm. We returned to the staging area and loaded onto the truck. Dower called Chin out and told him he would drive First Sgt. Gross. Walters, his former driver, was killed in the Tet assault. I got in the jeep with my radio.

We hit the main road and picked up speed. En route we see wrecked carts, trucks, vehicles, animals and debris scattered all about the roadway and alongside. We constantly swerve to avoid something in our path. The first checkpoint is like we left it, as I expected. I hope the bodies we bagged earlier have been removed. If not they are going to be very stinky. The next checkpoint is just as devastated. We don't stop until we reach the third checkpoint. This is our normal turn around spot and the first post manned by both U.S. and RVN MP's. We don't find any bodies but the stench of death permeates the area. Bodies are somewhere close. I wonder if we would have to go looking or if we were going to continue to Da Lat.

We move on. The streets are pockmarked so too the buildings. Evidence of intense firefights is everywhere. We see burned out ruins which were once the beautiful residences of the citizens here. The fragrant floral smells of this once attractive city are now replaced with the smells of decay and rot. What a tragic scene. We pull up to the Da Lat compound and see the massive wooden gates scorched and splintered. We enter the main courtyard and view the shattered windows and broken buildings.

"Damn," Rooster said as he walked up to our jeep, "I wonder if Brown made it through this?"

We assemble in front of the HQ office and are greeted by an unfamiliar First Sgt. Apparently the former went home DEROS (Deployment Ended Return from Overseas Service) or KIA (Killed in Action).

The new guy didn't waste time getting to the details. We will be retrieving bodies, simple as that. A number of civilian and RVN personnel are unaccounted for. We have a few GI's still not located also. The search would be methodical and all structures would be inspected for remains. We would begin with those directly outside of the compound and advance up the main roadway to city limits.

We split into five teams and started out. The first team went left, second team right, third team carries body bags, fourth team left, fifth team right. I was on the third team each consisted of four to six men. Before long we hear yelling and then somebody is throwing up.

"There are three here," someone from team two says. Team five found one and then team one discovered more. Fortunately for me, my team just passes out the body bags, we don't have to move or load the dead. Some are burned, others shot, and all are grotesque in death and repulsive to see. We spend the entire day on this detail with no breaks until 1500 hours. Fifty five, that's how many we found.

We washed up; trying to get the scent of bodies and rotting flesh off of us. No luck. Maybe it would dissipate as we return to Pre Line Mountain. We load up and head back, looking forward to a little rest and to forget about what we did today. We would return, if not tomorrow, soon, to do this same ghoulish task over.

The trip back is uneventful. As we turn into the compound, it struck me; I never thought this place would be such a welcome sight. Today, it is.

We spend two more days in Da Lat on the death detail. We don't find any missing Americans. Perhaps they were never there or were captured and taken away. I was just thankful to still be alive.

Days turned to weeks. By late February, the U.S. and RVN forces had retaken the major cities and areas that were under attack during the Tet Holiday. Stateside reports said we were handed our asses by the Valiant Freedom Fighters (VC). It was not difficult to comprehend whom the media favored in this conflict. I know first hand that the enemy forces were decimated, and heavy losses were suffered. Their success was short lived and it cost them mightily. That was not my concern. I still have several months left in this country. Old routines were reestablished, with infrequent firefights. Things slowly return to normalcy, if there is such a thing here.

On March tenth, a runner came to my hooch to report a call from the Red Cross. I ran to the Commo shack. It's a girl. She was born on March seventh. I am a Daddy!!

I ran over to the little PX next to the mess hall and bought a handful of cigars. I head back to the hooch and stuck a cigar in each guy's hand. They slapped me on the back, shook my hand and congratulated me on my good fortune. The elation lasted a lot longer than the cigar.

April has arrived and I go on another convoy to Da Lat. Rebuilding is underway both in the city and at the compound. The recovery period is slow. Trust needs to be renewed between the military and the population. We learn that the majority of the caskets we saw being transported and the funerals held at the churches were not actually burying bodies. The caskets were filled with guns, ammunition and explosives. On the evening of Tet the VC dug up the stashed weapons and launched their attack. This happened in all the major cities attacked.

Oh yeah, we found our friend Aaron Brown, the cook. He was on R&R in Hawaii during the Tet attack. Missed the whole party and got back just after the city was recaptured. His major complaint is that he has to clean up the mess.

We also got news from Gomer, who we left in Da Nang. He and Cherico and Lieutenant Derf all went to Hue, the ancient city where one of the most intense campaigns of Tet occurred. Seems Cherico and Lieutenant Derf are both up for a Bronze Star for Valor. They supposedly held off a Battalion of VC trying to overrun their site. They lost forty GIs at that location, thirty six wounded and four killed. Gomer is recovering from wounds he received. He is in a military hospital in Germany.

We loaded up supplies, mounted the trucks and start back to Pre Line Mountain. Once we unloaded everything; I head for the mail room. I can't wait to see some pictures of my kid and wife. I start my sorting routine, by postmark dates. The pictures were black and white Polaroid's. She has dark hair, and looks like she is all fingers and feet. Long arms and legs with a tiny little body attached. She weighed a little over seven pounds and nineteen inches long. It choked me up to know this is my child.

Still reflecting on the mail, I didn't hear Sgt. Dower come in. He stopped at my bunk to offer congratulations. Then he said, "You have been promoted again. Effective April 1st, you will be an E-5, Buck Sergeant!"

How about that? A Daddy in March and a Sergeant in April, think they will send me home in May? He laughed, "No, I don't think so, you just keep on hoping though."

We tested for Brown Belt in Tai Kwon Do the first week in May. Lieutenant Lee invited some of the big brass from the Korean Tai Kwon Do Federation to evaluate us. This was really a test for him because he told these folks that his students were ready. We wanted to look as impressive as possible. We had trained earnestly to perfect our patterns and moves for this event.

The Korean contingent came in by chopper and consisted of three Colonels and two Majors of the ROK (Republic of Korea) Army. We were totally rattled. We figured only a couple of people and not anyone of such high rank. These were field grade officers, straight from Na Trang. They were very congenial and impressed that several of us were from the 518th in Na Trang. Of course we didn't tell them our stay there was quite abbreviated. Each of us tested individually. We went through several stations and one or more of them observed our performance in each station.

We started the test at ten in the morning and ended around three in the afternoon. We did break for chow, 20 minutes and then right back into it. Talk about non committal, these guys didn't frown, smile, wince or cough, while we went through these exercises. When I did my flying leap kick to break the board, the evaluator stood on a bucket to see where my foot struck the board. Soaked with perspiration and really tired when it was over, we silently waited for the evaluations to be completed.

We lined up and bowed in respect to the evaluators, shook their hands and they left for the flight back to Na Trang. Lieutenant Lee told us the final validations would be sent to the Federation for ultimate decision on our elevation. He said we did well.

We got some new folks in since the Tet event. These guys are all new in country and very nervous about being here. Apparently they have been told to expect to die at the hands of the enemy. The first time we carried some of them out on a sweep; they were actually shaking and sweating.

Don't misunderstand, I was nervous too but I don't recall a fatal attitude on the mission. We have engaged in some limited firefights, but nothing like in the past. It seemed the

Tet Offensive as they are calling it knocked a lot of the fight out of the VC or decimated their ranks. Perhaps they were on a rebuilding program, or maybe just resting up for another big push.

Taking A Break

On 14 May 1968, I asked for a three day pass to go to Cam Rahn Bay. I had a tooth ache and needed a break from the monotony on the mountain. I grabbed a flight on a Huey going thru to Saigon. They would drop me off at the Cam Ran Bay R&R Center. On the flight we went over water which made me nervous. The fish didn't have any weapons, so no worries about hostile fire. The pilot spots some sharks and he decides to do some shark hunting. He hovered about fifty feet above the water and we shot at the sharks. Sounds cold hearted I guess, but if they had the opportunity we would be dinner. The activity ended soon enough, the co-pilot dropped his side arm (45 Cal. Colt) into the South China Sea. That will be an interesting report for sure.

I check into the billet and asked about a dental appointment. They scheduled me for the next morning. This place was just a vast sand box. The huts and buildings stretched for half a mile or more. One of the fellows I met from the American Division told me about a village where the guys assigned here lived. It was guarded by our GI's. Basically they rented a house with a woman. She cooked, cleaned and serviced the GI as long as he kept up the rent. When he left somebody else took over his hooch, his woman and any offspring resultant. This is condoned by our U.S. Military. I cannot believe it!

Cold drinks and good food are plentiful, as are sheets and circulating fans. Almost as good as the Air Force guys have it at their facilities. My dental work was swift, drill it out fill it with some kinda metallic filling and sent me off. I was done by ten thirty hours and looking for something to do. I went to the Exchange. I purchase a 35mm camera for my friend Ron. He and his wife have been great to Sherrie, my wife, while she was pregnant and all the time I have been away. His wife and mine are very close friends and hang out a lot. I also buy a Zippo cigarette lighter and have it engraved with the names of all the places I've been along with a map of South Vietnam.

I leave the exchange complex and head for the Air Transport office to see what time I can get a flight to Da Lat or the mountain site. I get booked for a standby on a caribou flight to Da Lat at ten o'clock the next morning. I felt confident that there will be a space for me. Most intelligent persons would never take a second flight in <u>that</u> aircraft, to <u>that airfield</u> and expect to survive the landing. As I walked out the door I hear Lieutenant Lee say, "Simpson, what you doing here?"

I reply, "I could ask you the same sir."

"Yeah, I am returning to my home and family. I must go through Nha Trang Headquarters first. I regret we did not get to the Black belt before I leave, but you can continue on when you return to your country."

I thank him for his time training us. He said his flight would leave in about a half hour. I shook his hand and offered him the Zippo lighter. He takes it reluctantly, "Thank you Simpson, I will keep this, it was an honor to have served with all of you."

He walks off, going home to his family, my opportunity is near. I return to the exchange for another Zippo, think I will have my name engraved on this one as well as the map.

The next morning that raggedy airplane was sitting on the runway waiting to shake, rattle and fly me back to my temporary Asian domicile. The flight takes a little over an hour and is just as exhilarating as the first one, nothing worked totally right, it shook and groaned and made me anxious to get off as soon as it stopped rolling. I did thank them for the ride (protocol and they outranked me anyway).

I hitch a ride with the fuel truck and get to the Da Lat compound just at meal time. After eating, I contact the site and let them know I am in Da Lat and will return with the next supply run. I go to the transient barracks and get myself a bunk and linens to make it up with. Three others are there, all heading for Pre Line Mountain. They are full of questions and keep me busy right up to supper. We walk to the mess hall and I visit with Brown, the Mess Sergeant. We talked about Wilson and when we each had last heard from or about him. He also tells me about his R&R and I brag about my daughter and, as any proud parent, displayed all the photos I have.

He suggests we meet at the club at 1900 hours. I tell him I will bring the new guys along. I didn't want to leave them alone. He was cool with that, so I went back to the barracks and asked them if they wanted to go to the club. We all went over and got a table and they started in with the questions once more. One guy is from Baltimore, the other two from La Crosse, Wisconsin. The one thing they had going for them was their MOS, Military Police. At least they had the training. I told them what we did as far as routines, guard duty, patrols, supply runs and occasional field intervention (set up ambush sites). They were worried about the M-16 reputation for jamming and carbon build up. I relayed my experiences with the weapon and so long as you keep it clean, there were no problems.

The major facts they needed to know were keep your weapon clean and close, same with your helmet and flak jacket. You could never have enough ammo and learn to select the best C-rations when the cartons were passed out. Otherwise you will eat a lot of lima beans and chopped meat (Ham, they tell us). I expect we would see a resupply convoy within the next couple of days.

Just about then Brown showed up and I introduced him to the new guys. It only took a few sentences until he was on the water buffalo attack in Da Nang and the newbie's yuked it up at my expense. I just smiled and let him go on. I would get my shot in when we talked about the Tet event and how much action he saw here in Da Lat. I was delighted to tell them Brown missed it all, except the clean up. I interrupted only to suggest that someone buy the next round.

We woke the next morning and the new guys were anxious to eat breakfast and look around the area. I suggested that we see if SSgt. Brown would take us around the city and tell us about what went on. They agreed excitedly. We head for the mess hall for chow and made our request.

Brown told us we got a new NCOIC for Pre Line, Gross was going back to the states. His tour ended. The new guy was on his second tour, he spent the first one in Saigon. At least he would be cooler on this one. Brown pointed him out and informed me he'd already told him the buffalo story. He was from New York, like Wilson and himself. I looked over toward the mess hall and saw a six foot tall black man chewing on a cigar and sizing up the troops as they walked by.

Brown introduced us. His name is Joe Mitchell, but everyone calls him Mitch. He turns to Anthon and asks, "Is he the one, the buffalo shooter?" Brown couldn't help himself and he just started laughing hilariously, joined by my newly introduced NCOIC. I stood there looking foolish. I asked if Brown told him that he paid the twenty bucks too, as he put as many rounds into that sucker as anyone else. I got a sobering look from Brown that told me he had left that part out. Mitch just looked over at him and started laughing again. I felt a little better that the story teller was now getting some of the grief.

At 1100 hours the convoy from Pre Line Mountain arrived. The newly promoted Staff Sergeant Dower is the convoy leader and his driver, the newly crowned highway ace, Sgt. Chin Lee. Introductions were made and we ate lunch prior to departing for the site. The trucks were loaded with provisions and fuel and made ready for the return trip. We said good bye to First Sgt. Gross. He was an excellent soldier and leader. I will miss his gruff manner and bellowing voice. Cincinnati, the guy who lobbed the grenade launcher round into the middle of the patrol brief, is headed back to the States with him. He just got promoted to PFC; he was an E-4 prior to the M-79 event.

After lunch we load up the troops. I ride back with the new guys on the ammo truck. Just to give them some excitement, I let them know what was under their backsides, after we were underway, of course. We rolled out and Chin leaned on the throttle. We were out of Da Lat city limits in less than five minutes and on the way to our little plot on the mountain. I tried to show the newbie's how to stay balanced, have their weapons ready and scan the area, while traveling on sub standard roadways, just under warp speed. They seemed to be doing a good job of adapting.

We rolled into the compound and stopped at the Operations Building. The CO came out to welcome First Sgt. Mitchell to the site and has his stuff moved to the NCO billet. We continued to the fuel depot, ammo bunker and the mess hall with the various goodies and after unloading everything, took the new guys to their hooch in our Sector (Lima).

It was a long few days for me and I was ready to kick back for a while, just relax on my bunk and read mail and think of going home. Rooster came in and wanted to talk, and then Chin. Finally I opened up a letter from my younger sister and read the news that her husband, Sal, was headed over here in the next month. He is ASA, (Army Security Agency), a cryptographer. I figured he would be in I Corps, probably Saigon. Those guys were kept so isolated; it was almost like being in prison.

I wrote her back and tried to give her some encouragement. He should be relatively safe, I try to assure her. Truth is rocket and mortar attacks are very worrisome, they were not very accurate, sometimes lucky; most times random strikes cause most of the casualties. Another letter from one of my other sisters tells me that my Mom has remarried a second time. Dad was still drowning his sorrow in vodka. He had a breakdown, stays in bed cry's a lot and won't eat. The other two kids are back with him; Mom doesn't have time for them. My Dad can't be doing much for any thing as far as caring for their needs. I hope my oldest brother is looking out for them; they definitely need someone at this time.

After supper I return to guard duty, with a small difference. I am now overseeing shift changes and escorting reliefs to the bunkers. My schedule was every four hours and Chin and Rooster were on the same schedule. This assignment was real sweet compared to what I used to do. I guess rank has some perks after all. I got the first shift and take the guys to their respective bunkers. We inserted the three new guys on our Lima group, each with an experienced soldier. There was some grumbling, understandably. Nobody wants to be with a rookie before his first firefight. When you are dependent on one another, trusting that he has your back is the most important thing. I assure them that things were just fine and reminded them that they too were once raw recruits and untested. I'm not sure it helped, but they understood things were not going to change, at least not tonight. I did my rotation and sat in the bunkers with the newbie's for about twenty minutes each. They seemed to be more at ease and their shifts went by with no problems, no shooting either.

The next few days went by with the normal activities, eat, sleep, patrol and guard duty. I began to teach the kid from Baltimore, Carson C. Carson, how to operate the PRC-45. He was a quick learner on the equipment, a little slow on the Zulu stuff, but his name was real easy to remember. I didn't ask if his middle name was Carson, I was certain his folks were a bit more creative than that! The two guys from Wisconsin were in their late teens and had never been anywhere besides their home state and Ft. Gordon, GA. As soon as AIT was completed they got thirty days leave and then onto an airplane to Saigon. All three seemed to be quick learners and I feel confident they will meld right into the operation. The smaller of the two, Ivan Jones was a farmer's son. The other kid; Steve Rowe was from a dairy farming family. A real experience for these two, ten thousand miles plus from home and stuck on a mountain top for the next eleven months of their lives, barring any unfortunate events.

Speaking of months, Rooster, Chin and I have less than two months left. I try not to think about it much though. I get too antsy and nervous.

Goodbye Pre Line

In today's mail, the last week of May, we received our certificates from the Korean Tai Kwon Do Federation. We were certified Brown Belts. Lieutenant Lee sent his congratulations and a little medallion for each of us. I wrote my wife that she better not mess with me, I have a Brown Belt. I knew her response would run something like, "You mess with me buster and you will get a belt from me." This came from a gal who was 5'4" and 105 pounds. In the mail I also got cookies and sweets in a care package and more pictures of my wife and baby girl.

The days turned to weeks of patrols and convoys and details and the ever present guard duty. It was quiet around here for several weeks and I got used to it. It did always make me a little apprehensive though, waiting for the next event to happen.

On 16 June1968, at 0030 hours, and I had just posted the second watch in the bunkers when I heard the familiar whoosh, thud and explosion. "Incoming," everyone shouted as we took cover. These rockets came at a fair pace, four, maybe five at a time and from different directions. I didn't hear any small arms fire which indicated that no one saw anything outside the perimeter. I slid into Lima One bunker and cranked the field phone. The Operations NCO answered and said to hold for a few and keep looking. The mortar pits were about ready for illumination flares.

Finally the flares went up and lit the sky in the brilliant reddish white glow. When they started their descent, I saw a few small figures darting just inside the wooded growth.

By the time we saw them they were back in the shadows. A few bursts from the M-60 let them know we saw them. We waited for their next move. More rockets hit, more flares sent skyward and more small weapons fired at the bad guys. A few small holes where the rockets hit, a resupply of the positions from where we fired on them and the night dragged on. The firing ended as abruptly as it had started.

Dawn broke and the newbie's were now veterans. We went to the mess hall and talked about the night's events. I wondered aloud to Chin if we acted the same way after our first time of actual engagement. He just laughed and said, "Remember the buffalo?"

July passed by fast with a couple of skirmishes, another rocket show and then it was August. I started preparing to go home, reunite with my family and a return to civilized society. SSgt. Dower extended his tour for six months. They promised him a fast track to E-7 and a choice of duty station when he returned to the US. He feels it is a wise career move for him especially since he not married and didn't have any kids. He hadn't heard from his parents in years. He only had contact with a sister in Wyoming. I never heard him talk about his family before.

The CO, now Captain Andrews was going home also. A significant number of us were clearing the compound at one time. A chopper would carry us from the site to Cam Ran Bay.

We all had the short timer complex. We were nervous, edgy and easily startled. We felt like the sky was falling and we might get smashed at any moment. It was real hard to shake the complex but we had to. It can get you hurt or killed.

Normally, short timers did not pull patrols or convoys and stayed on site. I concentrated on getting my stuff ready. I sold my sidearm to a Philippine contractor. I couldn't take it home with me anyway.

The new guys came along nicely and learned the ropes of survival in a remote area of the world. I was confident they would make it through and get back home. Rooster gave May, the dog, to one of the new guys from Wisconsin. He made him promise to keep her from getting eaten and gave him fifty bucks to care for her.

Before I knew it, there were only three days left before we would be on the way home. I packed my Tai Kwon Do outfits, along with the belts.

The next morning our housekeeper May gave us the balance of detergent, shoe polish and stuff provided her to do our cleaning. We told her to keep it, we wouldn't have room anyway. We also gave her twenty bucks each as a present. She was reluctant to take it, but after our insisting, she accepted. Rooster asked her to watch out for May and don't let the Montagnards eat her. She looked at him funny. We spent a few minutes explaining it was the dog May, not the girl May. We completed the next few days saying our goodbyes and exchanging addresses for stateside contact.

Finally we came to the last morning wake up and breakfast at Pre Line Mountain. My bag was packed and goodbyes have been said. As I stood at the doorway to the hooch, I recalled my first days here. It seemed like such a long time ago, and what an experience. I had probably said hello and goodbye to fifty other GI's, and it was finally me that gets to go home.

Carson C. Carson took my bunk space and I gave him the M-1 carbine I borrowed from the armory in Da Lat. It was just sitting around and feeling neglected anyway. I don't remember even firing a full magazine with it. This was during our body recovery period and I don't think anyone was around to claim it. You can't bring anything like that home so I figured he could use it, and maybe pass it along when he goes home.

The chopper landed at the helipad and we headed to get on board. It's a Chinook and it brought more replacements as well as some fresh supplies. We looked at the new guys as we passed and wished them the best. Captain Andrews greeted his replacement and passed command with a hand salute. "This is First Sgt. Mitchell," he tells the new CO, "he is your right arm and your most trusted ally. He and every man here will live or die for you. Treat them well and lead as you know how."

We climb on board, stow our gear and sit quietly. "Damn, never knew the CO felt that way, what a guy," I conclude. The door closes and we lift off the site. Goodbye Pre Line, it has been an adventure I don't want to relive.

We are headed for Cam Rahn Bay where we out process and get a few days to decompress. While there we enjoy certain luxuries like air conditioning, cold milk, beer and sodas, hot food, movies and real beds with sheets and pillows with pillowcases. It was a big change from down filled sleeping bags and wool blankets.

It took two days to out process, clear administrative procedures and receive orders and flight information. The last step was collecting our pay from the quartermaster which happens just prior to our boarding the plane. They figure we have less chance to get robbed or gamble it away. I report to the pay window, show my orders and identification. The clerk hands me a triplicate form to sign and initial in several places. I return it to him and he starts counting one hundred dollar bills, thirty three of them plus sixteen dollars and seventy three cents. He counted it three times then slides the bundle through the window and tells me to count it. I did with shaking hands and a nervous smile. I have never held so much money in my life.

We board the chartered jet and before long we are airborne and headed back to the United States. We will land at Seattle, check in and get new uniforms at Ft. Lewis. A bus ride back to the airport and we are homeward bound.

Leaving Seattle

It's 14 August 1968 and I am headed for the terminal, I'm going home! I got to the airport and called home, my flight stops in Minneapolis for fuel and from there to National Airport in DC. I will arrive about twenty one hundred hours tonight. I checked my duffel bag and sat down to wait for the boarding call. There are GI's all over the place, many like me, going home, others, shipping out for the Nam. You would think that after three hundred sixty one days, twelve hours and twenty seven minutes, I could patiently wait for the boarding notice.

That was not the way it was. Unable to read, nervously chain smoking, and biting my fingernails, I long to hear the speaker announce the flight. I look over the waiting room and see a lot of civilians waiting. Was I the only one in a military uniform? Some folks stared at me, or maybe I was just imagining it.

They called for boarding for Flight 459 for National Airport at Gate 3A. I jumped up as do all the others waiting. I forget there was a seating protocol. I've never understood why the First Class passengers get on ahead of Coach. Maybe it's so they can scowl at the poor folks as they pass by. Finally, my row was called and I step briskly toward the plane. The Stewardess directs me to the rear section of the aircraft. It has more room than the Caribou and a nicer interior for sure. I have an aisle seat and settle in for the trip.

I hear a male voice calling for the Stewardess and vocalizing about some criminal on the plane. I look back for someone in handcuffs or something and then I realize he's talking about, "baby killers, war criminals and murderers." He's talking about me. He was actually pointing at me and telling the lady he doesn't want me on this plane with decent human beings. A few others started to echo his comments and others yelled at them to shut up and mind their own business. The Stewardess walked up to my seat and asked me to come with her. I was embarrassed by the commotion, hurt by the meanness and vitriol expressed by my countrymen and concerned that she was going to toss me off the plane. "I just want to get home lady, please, have a heart."

We exited the coach area and she said, "Don't pay any attention to that guy and his cronies. You are getting a free upgrade to First Class, my treat. Now what would you like to drink? After we took off, she returned to say that the guy thinks I got booted from the flight. I wondered what he would think when I get to DC the same time he does. I really enjoyed the flight home, the meal was great and that seat, it was so comfortable!

We land and taxi to the terminal, the engines shut down and the door opens, the pilot gives us permission to deplane. I file off and thank the stewardess. She smiled, "Oh, Sgt. Simpson, my husband is on his second tour there. He left two months ago, assigned to some little mountain outpost just outside of a place called Da Lat. I know what you have been through; I just hope you won't be angry about what you've seen on this plane. A lot of people, most of us in fact, do not feel that way."

I looked at her name tag, Valerie Mitchell. I start to say something, but just stammer a silly, "Thank you," and walked away. It is cliché but true, it is a small world.

The Homecoming

I get to baggage claim, retrieve my duffle bag and then head for the main terminal area. I choke back a tear as I see Sherrie coming down the terminal hallway to great me. I was home in one piece and reunited with those I held dear. I smiled, we touched, embraced, kissed, and said, "I love you," almost in unison. We found her father who is waiting to drive us home, we shake and he says "Welcome Home", that is possibly the longest sentence he has ever spoken to me. I smile and say "Thanks". We sat in the back seat, unwilling to let go of one another, giddy and smiling.

Once we get home my mother-in-law stands at the door. She hugs me and tears streamed down her cheeks. She simply says, "We missed you."

I answer trying to make light of the tender moment, "I missed you all too. Thanks for the spinach", she chuckles. We walk to the bedroom and I look into the crib where this tiny little person was fast asleep. What a beautiful child.

I have thirty days to reacquaint myself with family and friends, and then its back to the
army life style. I still had thirteen months left. I would finish my career as a soldier at Ft. Meade, Maryland.

After several days my baby girl begins to recognize me. She even smiled at me. Old friends are stopping by to visit. It was sort of like an opportunity to be reviewed and evaluated to determine if it was safe to be around me. After all, I am a "warfare trauma veteran," prone to go berserk at any moment. I think they figure it to be safer at my in laws, than me visit them at their homes. Don't mistake my comment here, not all my friends are like this. They loved me, they were just trying to confirm or deny all they read and heard. I must admit that if they were around when I watched the national news broadcasts, my reactions would give cause for them to wonder.

We move to an apartment in Laurel, Maryland about six miles from Ft. George G. Meade. While there we visit Ft. Meade to locate the commissary and other facilities. I also visit the 414th Signal Company, my new assignment. I report to the 414th and am assigned duties as the Company mail clerk. This was a real gravy position, no formations, weekend duties and only infrequent details as week night CQ, (Charge of Quarters).

After about six months of serving with this Company, I apply for and am reassigned to the Installation Fire Department (The military participates in a program called Project Transition, where those within six months of discharge could train or re-visit career fields as they return to civilian life). I work out of a Military Fire Station as a crew chief. This goes well until the "C" Shift turns our fire truck over in front of the Officers Club. It just happened to be the Sunday that the Installation Commander was having his monthly breakfast for staff. By the following morning the military station had been closed and all except me and another guy (He was from the Atlanta Fire Department) had been reassigned. The two of us were detailed to the Tipton Army Airfield and worked with the civilian firefighters at the airfield. We were welcomed and honored to be in their company. This is where I will spend the balance of my Army career

I am discharged from the U.S. Army in October of 1969 some three years and five days have passed. I have seen more than a lifetime of events, adventures and experiences that I never anticipated would be part of my life. I subsequently returned to my career as a firefighter in Washington, DC, finally able to return and fulfill my life long ambition.

EPILOGUE

I am obviously older now and hopefully a little wiser. My war experience is now a mix of memories good and bad. I feel the compulsion to relate my story to respond to questions asked over time by both friends and my family. I remember many details of my experiences, especially those where fear was a formidable concern. I can assure you that my life was affected by the war. I experienced the brutal nature of mankind and the ugliness of war. There are still those mind pictures of the dead and dying, their screams of anguish and pain, the horror of mangled and broken bodies. Many adjusted to civilian life with minimal discomfort. Others have had a greater difficulty in assimilating into society. That is a tragic commentary of our own societal response to addressing the need for healing.

It took me a several months to not react to loud noises, longer still to overcome some disquieting moments and control sudden urges to be aggressive when angered. But it is true that time heals wounds, both the physical and the psychological. The love and nurturing of friends and family make the transition easier though.

I have visited the Vietnam Memorial twice since it was completed. My first visit was the late eighties. I was assigned to the Fire Prevention Bureau and while on the street I stopped to see this "Monument to Hero's" who were now a part of me. I felt uncomfortable standing there. I am overwhelmed by the grief and guilt I feel. I stayed only a few minutes and went back to my car. The second visit was the same. I walked along the dark cold slabs of stone, etched with names, some of whom I recognized. I grieve because of the sacrifice of their lives. I'm guilty because I lived through it and they did not. I am angry that they have been vilified by those they died to protect. Those with no regard for the freedoms they enjoy who criticize and slander, the very folks who sacrificed their lives to keep them safe in their own little world.

I know names of some on that marble monument. I shared a small part of my life with them. They were Americans, all colors, ethnic groups, privileged and poor, draftees as well as volunteers. I can assure you they loved life and wanted to continue with it. The choice was not theirs to make. We will never know what potential greatness was lost by their sacrifice. They were future leaders, fathers and mentors, whose giftedness will never be realized. As a nation I feel certain we have suffered for their sacrifice.

I am telling my story, not for recognition or reward. My purpose is to relate the ugliness of humanity and tragic results of letting politics dictate tactics. Vietnam was not a military loss for the United States. I think of it as a time when we lost our vision and our soul.

This year my wife Sherrie and I will celebrate forty four years of marriage, we have three children who I am very proud of. I have grandchildren too. This story is my attempt to tell them that the "American Dream", is real and that there is a cost. I pray the time will never come that we as a family will cease to appreciate the hard fought freedoms we have. I would also hope that they will always have an opportunity to live in the "Land of the Free and the Home of the Brave".